PHYSICS RESEARCH AND TECHNOLOGY

FROM ELEMENTARY PARTICLES TO THE LIMITS OF THE INFINITE COSMOS

Physics Research and Technology

Additional books and e-books in this series can be found on Nova's website under the Series tab.

Space Science Exploration and Policies

Additional books and e-books in this series can be found on Nova's website under the Series tab.

PHYSICS RESEARCH AND TECHNOLOGY

FROM ELEMENTARY PARTICLES TO THE LIMITS OF THE INFINITE COSMOS

VAGGELIS TALIOS

Copyright © 2020 by Nova Science Publishers, Inc.

All rights reserved. No part of this book may be reproduced, stored in a retrieval system or transmitted in any form or by any means: electronic, electrostatic, magnetic, tape, mechanical photocopying, recording or otherwise without the written permission of the Publisher.

We have partnered with Copyright Clearance Center to make it easy for you to obtain permissions to reuse content from this publication. Simply navigate to this publication's page on Nova's website and locate the "Get Permission" button below the title description. This button is linked directly to the title's permission page on copyright.com. Alternatively, you can visit copyright.com and search by title, ISBN, or ISSN.

For further questions about using the service on copyright.com, please contact:
Copyright Clearance Center
Phone: +1-(978) 750-8400   Fax: +1-(978) 750-4470   E-mail: info@copyright.com.

NOTICE TO THE READER

The Publisher has taken reasonable care in the preparation of this book, but makes no expressed or implied warranty of any kind and assumes no responsibility for any errors or omissions. No liability is assumed for incidental or consequential damages in connection with or arising out of information contained in this book. The Publisher shall not be liable for any special, consequential, or exemplary damages resulting, in whole or in part, from the readers' use of, or reliance upon, this material. Any parts of this book based on government reports are so indicated and copyright is claimed for those parts to the extent applicable to compilations of such works.

Independent verification should be sought for any data, advice or recommendations contained in this book. In addition, no responsibility is assumed by the Publisher for any injury and/or damage to persons or property arising from any methods, products, instructions, ideas or otherwise contained in this publication.

This publication is designed to provide accurate and authoritative information with regard to the subject matter covered herein. It is sold with the clear understanding that the Publisher is not engaged in rendering legal or any other professional services. If legal or any other expert assistance is required, the services of a competent person should be sought. FROM A DECLARATION OF PARTICIPANTS JOINTLY ADOPTED BY A COMMITTEE OF THE AMERICAN BAR ASSOCIATION AND A COMMITTEE OF PUBLISHERS.

Additional color graphics may be available in the e-book version of this book.

Library of Congress Cataloging-in-Publication Data

ISBN: 978-1-53617-456-4

Published by Nova Science Publishers, Inc. † New York

To my family;
my wife Lena
my daughter Stella
my son-in-law Steff
and my granddaughters
Isabella and Evelina

Our up-to-date knowledge of the Universe
and a new suggestion of the creation of the Cosmos
according to the *"Chain Reaction Theory"*

Contents

Prologue		xi
Some Words about the Theory of the Chain Reaction		xix
Introduction		xxiii
Chapter 1	General Facts	1
Chapter 2	The Evolution of the Human Knowledge about the Universe	9
Chapter 3	The Universe as It Is Known Today	37
Chapter 4	Matter, Elementary Particles and Fundamental Forces	55
Chapter 5	The Cosmological Theories and the Theory of the "Big Bang"	71
Chapter 6	The Theory of the Chain Reaction: A New Attempt to Explain the Creation and the Operation of the Cosmos	81

Chapter 7	The Creation of Our Universe the other Universes the Antiuniverses and the Cosmos: The Separation of Matter and Antimatter. The Cosmos in Its Actual Form According to the Theory of the Chain Reaction	109
Chapter 8	A Brief, Imaginary, Theoretical Journey, to an Antiuniverse	121
Chapter 9	The Fundamental Unanswered Questions about the Creation: The Answers of the Theory of the Chain Reaction	137
Chapter 10	The Theory of the Chain Reaction the Trilogy of the Creation and the Unification of the Fundamental Forces and Physical Theories	145

Epilogue   159

Glossary and a Brief Explanation of Several Basic Terms Used in This Book   161

References   169

Author Contact Information   171

Index   173

PROLOGUE

Dear friend, reader,

The book, *From Elementary Particles to the Limits of the Infinite Cosmos*, is the first of a series of three works, described in respective three books, which, after being completed, will form the "trilogy of the creation". With the "trilogy of the creation", I make an effort to explain the physical phenomena related to the creation, the existence and the functioning of the material part, of our Universe and the Cosmos. Thus, starting from the elementary particles, "pointons" and "antipointons", –as I define and describe them in the "theory of the chain reaction"–, passing through the atoms, molecules, matter and antimatter, we find the Stars, the Solar systems and the Galaxies, the Universes and the Antiuniverses, cross the immense, absolutely void spaces, in order to reach, finally, the ends of the infinite Cosmos.

In this book, I describe the human perception of the Universe and the Cosmos, as we know them today, the existing cosmological ideas about the creation and I suggest the "theory of the chain reaction", for the completion or replacement of the most prevailing theories until now. The reason for which I suggest the "theory of the chain reaction" is that I believe it explains the formation and the evolution of the Universe in a more convincing way. On the contrary, I feel that the most prevailing theories, as we are going to discuss, are based on disputable axioms, leave a lot of questions unanswered and from one point, they are not able to describe the physical phenomena related to the creation with at least some rudimentary or reasonable credibility.

The second and third books of the trilogy correspond to two independent works with theoretical details. The second book describes the theory about the "unification of the fundamental forces and physical theories" which is one of the most basic issues concerning the creation and holds great interest for the modern scientists. The third book describes the theory of the "formation of matter and antimatter", an original work that explains how matter, antimatter, material bodies and all the interactions, occurred from the elementary particles, point charges, "pointons" and "antipointons"; –two particles without mass and dimensions– and from just one interaction; the "electromagnetic interaction". These two books could be considered in a way the basic theoretical support for the "theory of the chain reaction" and the general complement to the "trilogy of the creation".

Of these two theories, the theory of the "unification of the fundamental forces and physical theories" is suggested for the replacement or completion of the corresponding relevant theories that treat the unification. The next theory, the theory of the "formation of matter and antimatter" is suggested for the partial completion of the theory of "quanta" in what concerns the part of the theory describing the interactions between the particles of matter –the fermions– and the respective particles, of the interactions –the bosons.

Perhaps the title, "trilogy of the creation" might seem somewhat extreme for the whole work. This is why I would like to make clear from the beginning that this work does not aim to introduce a scientific textbook that will assemble all our theoretical and experimental knowledge about the creation, but to place the real foundation of a structure on which we shall be able to base and incorporate clearly and successfully all our up-to-date theoretical and experimental knowledge on the creation.

Although in many cases this book invokes the theory of the "unification of the fundamental forces and physical theories" and the theory of the "formation of matter and antimatter", it is essentially a self-contained and independent textbook, addressed to all those who have elementary scientific knowledge and are interested in being updated about issues relevant to the creation and the function of the "microcosm", the "megacosm" and the "cosmos of the infinite" in the Universe and the whole Cosmos. It is also addressed to young scientists who probably have not yet had the chance to study or give a lot of thought to questions, for example what matter is and how it was formed? What and where antimatter is? Whether the Antiuniverse exists or not? What and where it is? Whether other Universes and Antiuniverses exist? What is the cause that makes the material bodies be attracted between them? Whether the events and the various physical phenomena in the Cosmos follow a random course of evolution, or they are guided by an "Upper Force"? etc.

In the "theory of the chain reaction", described in chapters six and seven in this book, I introduces with complete reasoning, several new theoretical ideas such as: the ideas about the properties and the behavior of the elementary particles, –where the formation of matter starts from entities, particles-charges, with neither mass nor dimensions, which however have inertia and move with finite speed–, the ideas about the structure of the atoms and the cohesion of the nuclei, –where we establish the rotating orbits of the particles, inside the nucleus too–, the ideas about the formation of matter and antimatter and about the unification of all the fundamental forces that exist in the Cosmos, in levels of only one interaction, the "electromagnetic interaction". Farther more the concepts of the "cosmogonic gas", "existing antimatter", "Antiuniverse" and multiple "Universes" and "Antiuniverses" from which the "Cosmos" is constituted, as we describe it analytically in chapters six and seven, respectively, are also introduced for the first time.

The other two books describing the theory of the "unification of the fundamental forces and the physical theories" and the theory of the "formation of the matter and antimatter" are independent works, which, as mentioned above, supplement the "theory of the chain reaction". These two books are addressed to those interested in reading the theoretical details about the unification of the fundamental forces and the physical theories and how, according to the "theory of the chain reaction", matter, antimatter, masses and material bodies were formed from pointons and antipointons.

Although, in general, I tried to use the simplest scientific terms and remarks possible, in the description of the various events and physical phenomena, I considered it was worth, including a small glossary at the end of the book, with brief explanations of the most basic terms I used. Thus, the reader who does not know an eventual term will be able to find it in the glossary, instead of looking it up in other books, in order to resolve his/her questions.

The more important of the motives that lead me to write this book, as well as, the whole the "trilogy of the creation", are many. From those motives, the following can be distinguished:

a) I have read a lot of textbooks in cosmology, physics, astrophysics, astronomy and mathematics and studied many papers related to the issues of the creation. However, I was not able to find any convincing answers that would cover at least a small part of the multiple questions I had; some of which I have already mentioned above.

There are abundant textbooks referring to the Universe, as well as equally abundant papers describing the megacosm and the microcosm. However, I have not been able to find one complete book that describes the evolution and the advances made until now on those subjects in a convincing, responsible, simple and understandable manner. The relevant cosmological theories, the most prevailing of which is the theory of the "big bang", that is theoretically supported by the theories of "relativity" and the theory of "quanta", which are supposed to explain the phenomena of the creation and explain the evolution of the Universe, besides being based on disputable axioms, are also covered by an incomprehensible veil of mystery where almost nobody can pass through.

The two phrases below are characteristic: one is from Bert Russell's book, the "ABC of relativity": "It is true that there are

Prologue

innumerable papers on the theory of relativity, but they generally stop being understandable at the point that they start to say something significant" and another phrase from St. Hawking's "A brief history of time": "When someone asked an eminent scientist if it is true that only three scientists can understand the theories of the relativity, he remained thoughtful for a while and then answered that he could not think who the other two scientists who understand the theories are?!!!"

b) The second and more basic motive that led me to write this book was that together with my effort to find some convincing answers to my questions, I studied and completed some of my own scientific ideas, which, as weird as it may seem, explain many of the above issues more convincingly. I considered I had to make these ideas known. However, as the project of its disclosure advanced, these ideas resulted in three interconnected theories: the "theory of the chain reaction", which I describe in this book; the theory of the "unification of the fundamental forces and the physical theories" which I describe in the book, "The Real Grand Unification", which has recently been published and the theory of the "formation of matter and antimatter", which I shall describe in the third coming book of the "trilogy of the creation".

c) About sixty years ago, I had long discussions about several of my initial ideas with the ever memorable professor of mine, Th. Kougioumtzelis, who had advised me at that time to study them more attentively and, after having completed them, to publish them.

d) At the same time, however, wishing to offer something more substantial and useful to the readers and the young scholars, who may eventually not agree with the ideas of the "theory of the chain reaction" and in order to be more correct in the comparison of my ideas and those of the other theories and after having studied a lot on the whole subject, I considered it right to write as briefly as possible but completely and understandably the perceptions that exist until now. I believe that in this way the readers will have the chance and the advantage to be informed about the existing theories and then proceed to their comparisons and reach their own final conclusions unbiased.

While writing this work, I had defined as my basic goal, to describe and explain the various physical phenomena in a way as simple and understandable as possible, which at the same time would also be a scientific way. In parallel, I tried to provide answers to the existing questions; answers that would be clearer and more convincing and understandable, than the existing ones, if consider that there are answers indeed.

So, I used the simple Euclidian mathematical reasoning and I did not need to establish any notions as "space-time", "space of four, five or six, etc., dimensions", "length contraction" or "time dilation", "undefinability", "virtual particles" etc., notions that I do not discard but simply, I could not find a clear description for them in order to be able to study them and then judge whether I should accept or reject them. In any case, what I wish the readers to know is, that notions such as the above, not only have not been used, but have neither been necessary for the description of the "theory of the chain reaction" or then, for the completion of the writing of the "trilogy of the creation". Thus, I avoided as much as possible the complex, unreasonable and unsupported thinking in the descriptions of the events and I did not establish any axioms, laws or notions that I have not founded.

In general, I followed the example of the Ancient Greeks, who, with primitive instruments and using simple but logical scientific thinking, as I describe in the second chapter, calculated the size of the Earth, the Sun and the Moon and measured the distances between the Earth and the Moon, the Earth and the Sun, etc., in our Solar system; with that data they founded the first concepts about the structure of our Universe, concepts that remain valid until now. Following this same strategy in the "theory of the chain reaction", I believe I managed to give the reader the feeling that he/she reads a simple text book –that describes the creation of the Cosmos from the elementary particles to infinity– and not a complex scientific study.

Prologue xvii

Nothing written in this book or in the whole trilogy aims to change or invalidate the existing perceptions about the creation; they aim to supplement them and clarify some parts of them. Concluding the prologue, I would like to emphasize that both, the work in this book and the whole work of the "trilogy of the creation" are theoretical works of purely scientific nature and have nothing to do with Religious or Life matters. They concern only the material part of issues related to the formation of matter, antimatter, the material bodies, the Universe and the material Cosmos.

I would like to thank in advance all those who will read the book and I declare that its publication, does not aim to a personal profit or any personal ambitions of mine, but it is done because I feel obliged to disclose some thoughts and some innovative ideas about those issues.

SOME WORDS ABOUT THE THEORY OF THE CHAIN REACTION

Vaggelis Talios, a researcher and thinker, inventor of the "theory of the chain reaction", after a short comprehensive summary of the data and ideas that science accepts up to now about the existence and the evolution of the Universe and the Cosmos, describes a new, personal idea of his, about the creation. According to Vaggelis' idea, "... In the beginning the Cosmos consisted of a vast, absolute void of infinite size. At a certain moment, an abnormality occurred and an elementary particle-charge was formed with neither *mass* nor *size*. The theory named this particle *'pointon'*. In order to restore this abnormality, opposite particles to the first one, the *'antipointons'* were formed. The formation of the antipointons was considered another abnormality and new pointons were formed in order to restore it. Thus, a flash, chain reaction started, producing pointons and antipointons; this reaction is still going on, until now, at the limits of the Universes and the Antiuniverses. The production of pointons and antipointons occurred and occurs without energy consumption, as, in any case, there was no energy because, according to the "theory of the chain reaction" the Cosmos started from 'zero'.

"A characteristic property of the pointons and antipointons created was that the opposite particles were attracted and the identical ones were repulsed. The theory named this attraction or repulsion, which corresponds to the actual electromagnetic force, 'electromagnetic interaction'. The electromagnetic interaction was the cause that sometimes, led the pointons to capture antipointons –or vice versa– and force them to spin around them. So the 'quarks' up and down, the 'antiquarks', the 'electrons' and the 'antielectrons' were formed. These were charged particles and, contrary to the pointons and antipointons they had, *'mass'* and *'size'*.

"Then, from the quarks and antiquarks, the 'protons' and 'neutrons' were formed, as well as a branch of the 'electromagnetic force', the 'strong nuclear force', which joined protons and neutrons to form the nuclei of the atoms. From the nuclei and the electrons, 'atoms' and 'antiatoms' were formed. Here, always according to Vaggelis Talios' idea, *"the situation was reversed"*, as a second branch of the electromagnetic interaction, the 'gravitation' occurred between atoms and antiatoms, as a gravitational force, with the atoms attracting similar other atoms and repulsing antiatoms and antiatoms attracting similar other antiatoms.

"This reversal was the cause that resulted in the creation of matter, anti-matter and then the Stars, the Solar systems, the Galaxies, the Universe, the Universes, the Antiuniverses and the Cosmos…"

This is very briefly the "theory of the chain reaction", which, according to Vaggelis, explains the creation of the Cosmos in a more reasonable and convincing manner, than the other cosmological theories. However, you are going to judge this after having read the theory.

ACKNOWLEDGMENTS

I proceeded to the first –trial– edition of this work, by the book with the initial title "The Cosmos, the Universe and the Antiuniverse" in Greek language in order to disclose some personal innovative thoughts of mine about the issues of the creation, of the Universe and the Cosmos, to my close friends and scientists.

By no means had I thought that the work I had published, might take the form of a complete theory about the creation, making it able to be compared to the most prevailing cosmological theories and aim to complete or to replace them.

However, the warm welcome and the serious interest of my friends, expressed in favorable comments and interesting suggestions, gave me the impression and made me feel that I should review the formulation of the text of this work, –in order to make it as more complete, interesting, responsible and understandable as possible– and publish it, with its new definitive title: "From the inside of quarks and up to beyond the Universe".

In the sentence above, the phrase –and publish it, with its new definitive title–, was written rather out of date. As a novice writer then, –about ten years ago–, I did not foresee that in physics and cosmology nothing is definite and that a book containing a new fundamental theory of the creation is constantly improving. These improvements led me to the new revised version and new title, "From Elementary Particles to the Limits of the Infinite Cosmos".

In this very brief note, I would like to emphasize, for ones again, the unreserved assistance I enjoyed from my publisher and my friends, in all this effort of mine … and thanks them.

INTRODUCTION

In sciences and especially in physics and in mathematics, arise questions, to which sometimes, no correct or convincing answer has yet been given. These questions between physics and mathematics have one essential difference. Whereas for the questions in mathematics we often do not know whether a correct answer exists, for those in physics, most of the time, we know in advance the result of the answer; but the answer itself however, has been lost in the past and we try to find it today.

So, it is not enough for physics, as a science, just to find the correct result to a question, as we already know this result, but it is expected to discover the correct answer too. However, the fact that we are already aware of the result of the questions in physics, often we do mistakes in the answers given. So, while we believe that we have found the correct answer to several questions, we then discover that this answer is wrong. Moreover, this is the most important reason that leads to the establishment of a theory, for several years, or decades, or even centuries and then at a certain time, we find out that this theory is not correct.

A typical example of erroneous interpretation is the explanation concerning the reason for which the various material bodies are attracted by the Earth. It had been said that this phenomenon, is due to the fact that Earth is the center of the Universe and as all bodies are attracted to the center of the Universe, they are consequently attracted to the center of the

Earth. This interpretation constituted the theory that established Claudius Ptolemy's geocentric system, which as we shall see below, even though wrong, prevailed in the World for a period longer than fifteen centuries.

Today we know that the correct answer of the attraction of the bodies to the center of the Earth is that the centers of gravity of all bodies, the Earth included, are attracted between them and create this phenomenon. Copernicus' theory of the Heliocentric system, –according to which the Earth moves around the Sun and not the Sun around the Earth–, is based on the interpretation given by Newton with the law of the universal attraction. The whole dynamics of the Galaxies and the function of our Universe and the Cosmos were based later, on the same interpretation.

However, although the mechanism of the attraction of material bodies has been studied and the correct answer for the reason why the material bodies are attracted to the center of the Earth has been given, the cause of this attraction between material bodies has not been found yet. Many theories have been suggested about the cause of this attraction, but none has been able yet to convince scientists about its correctness. Therefore, in this case too, it is absolutely certain that among the many ideas suggested only one will be probably proven correct. The others will be rejected as erroneous ideas.

Another case of erroneous interpretation of a physical phenomenon was that after the discovery of the three particles, electron, proton and neutron, scientists considered that all the elementary particles of matter had been discovered. These elementary particles, with the electromagnetic and gravitational forces and the particles of their interaction, photons and gravitons, which are also the carriers of these interactions respectively, were considered to be all the elementary components of matter. Indeed, several scientists believed that the research of the microcosm ended at that point. Today, however, after the discovery of the quarks, the weak and the strong nuclear forces and their carrier particles, it has been found that the fundamental components of matter do not end at the above particles. So, the discovery of the new particles questioned again the previous ideas, which have already been abandoned. Today, a new theory is sought, which will incorporate and unify all new scientific data about the composition, the structure and the functioning of the microcosm.

Having the above in mind, when I once was reading a book about the Universe I paused at a point where its author explained that: "The vastness of the Universe constitutes a valuable source of physical knowledge…" and then he wrote, "…however, the Universe, despite the scientific explanations provided by Physicists, Astronomers and Mathematicians, continues to excite the imagination of humans…" etc.

The reason I paused at that point was that I wondered about what vastness of the Universe and which scientific explanations the author of the book I read meant as:

I had rather opposite views about the vastness of the Universe, as I describe them later on; at that moment I summarized those views in the following phrase: "however big the size of our Universe, it does not reach one thousandth of a thousandth the total size of the rest of the Cosmos and the existing unutilized and absolutely void space".

Now, in what concerns the scientific explanations mentioned by the author of the book, I wondered whether they explanations answered at least a minimum of the big questions I had about certain physical phenomena related to the creation of the Universe and, particularly, the following questions:

Where may those vast quantities of energy that exist today in the Universe have been found?

From what and how was the matter formed? Was it formed out of nothing or not?

If matter was not formed out of nothing, then where and how were the substances that formed it, found?

Is there the antimatter or not? What is the antimatter and, in case it exists, where might it be?

What is the cause that makes the masses attracted?

Why are the masses attracted, whereas they should reasonably be repulsed since they consist of homonym elements?

Why is the law of the universal attraction inconsistent in very short ranges? –According to the above law, the force of attraction between masses in short ranges should tend to infinity. On the contrary, this force in the real conditions in nature is eliminated–.

Are there masses that are repulsed?

When did time begin and when will it end?

Where does space end? Where does the Universe end? And where does the Cosmos end?

Do the events inside the Cosmos and the Universe develop randomly or are they guided by a "Higher Force" which we do not know?

I had studied a lot of textbooks in an effort to find some convincing answers to the above questions. But I had found out that: "the more I broadened my knowledge the more my questions broadened and became more perplexed".

For example, I mention that many theories have been suggested about the structure and the development of the Universe, as the theory of the "big bang", the "steady state theory", the "theory of the continuous dilation", the "theory of the perfect cosmological principle", the "plasma theory" etc., but none offers a convincing proof about its correctness.

In this case, then, and after having studied the above theories, I felt somewhat embarrassed finding out that instead of having a convincing answer about the creation of the Universe, I had one more question. As it is not possible for all the above theories to coexist, because normally there is only one real way in which the Universe was formed, then which of the existing theories is the one that I should accept? And this, because it is normal, that I need to accept just one of all those theories.

I had already questioned much the selection of the theoretical aspect with which I would agree most. However, all the theories I had read contained a lot of unclear and vulnerable points; even the theory of the "big bang" that expressed the latest views about the formation and the development of the Universe was not an independent and complete theory, as it was based on disputable axioms and a lot of inaccurate assumptions. For instance, the theory could not explain where and how that tiny ball, – the "cosmic egg", as it is usually called by several authors–, with the infinite energy, infinite density and infinite temperature, from the explosion of which the creation started; was found. Then, the theory of the "big bang" could not explain how the above energy that the small ball contained was transformed into matter. Also, the theory did not provide an answer to several other basic questions as I describe in the relevant chapter.

However, further than the above unclarities, the theory of the "big bang" introduces many other obscure points in order to conclude to the formation of the first elementary particles and then to the formation of nuclei, atoms, molecules and matter. At the same time, we cannot consider the views of the theory about the fate –in fact about the disappearance– of the antimatter, convincing.

All the above were the causes that led me to work on and study in depth the data that science provides until now, try to examine as closely as possible the issue and think of new theoretical ideas that gave me more convincing answers than the existing ones. The result of this laborious and long –about twenty years– effort of mine, is the writing of this book, the main subject of which is the description of the "theory of the chain reaction", which I believe explains quite convincingly the development and the formation of the Universe, the other Universes and Antiuniverses and the Cosmos. However, I leave you to judge this after having finished reading the book.

Chapter 1

GENERAL FACTS

"However, if we discover a complete unified theory, its general principle will soon be understood by everybody and not only by physicists. Then we shall be all together, philosophers, physicists, scientists and everyday people, be able to participate in a discussion about why it happens that the Universe and us to exist."

From Stephen Hawking's *A Brief History of Time*

This is the ending paragraph of Stephen Hawking's book *A Brief History of Time*. I chose this same paragraph as the beginning of the writing of the "Trilogy of the Creation" aiming to emphasize the simplicity of thought that I tried to give to the whole content of the trilogy. At the same time, however, I would like to make a point, of the fact that it is rather a deception to believe that today's established situation, with all those taboos, that it has created around itself, will accept a simple and understandable theory, that will enable simple people to participate in a discussion about the Universe.

The writer's point of view

THE UNIVERSE THE OTHER UNIVERSES THE ANTIUNIVERSES AND THE COSMOS

When we use the word "Universe" we mean that part of space where Stars, Solar systems, Galaxies and voids among Stars, Solar systems and Galaxies are formed, develop and coexist.

However often, instead of saying Universe we use the word Cosmos meaning the multitude of celestial bodies as a whole of harmony and order. Until now cosmology, astronomy and the science of physics have not yet determined a clear concept for the word Cosmos. Some identify the word Cosmos with the Earth and many times with the sky and the Universe. Others distinguish the material from the vital Universe and when using the word Cosmos they refer to the whole of the material and vital Universe at the same time, etc.

However, as we shall see in the respective chapters of the "theory of the chain reaction", together with the creation of the Universe, the "Antiuniverse" is also created. It is also very possible that besides our Universe and Anti-universe several other Universes and Antiuniverses were or are created, with the same, similar or completely different characteristics than the ones of our Universe.

Vast "void" spaces exist between the Universes and the Antiuniverses. It is also certain that immense "absolutely void" spaces –spaces without any interactions, which have not been used yet– with a size that exceeds any possible limit of our thought or imagination also exist outside the Universes and the Antiuniverses. These spaces extend to infinity. We shall include in the word "Cosmos" all those "Universes", "Antiuniverses", the infinite "void" and "absolutely void" spaces, together with all the events that develop inside those spaces, as well as anything else that exists. This will be the meaning of the word "Cosmos" when used it in this book, as well as in all the three books of the "trilogy of the creation".

The Obscure "Notions"
the Initial Elements of the Creation

Initially, there was absolutely nothing in the Cosmos. Neither light nor dark, neither cold nor heat, nor energy and not matter; the absolute zero prevailed everywhere. The only elements constituted the Cosmos except the absolutely void space were the infinity "concepts", which, however, were in a certain latent condition as there were no respective events upon to which these concepts could be manifested.

For instance, the concepts of interaction and force, attraction and repulsion existed, but they were in silence as the corresponding masses or the appropriate charges, on which forces would be exerted in order to create the attraction or repulsion, did not exist. The concepts of the straight and curved line and the circle existed, but not the respective particles or points that would allow the formation of those concepts existed. The concept of size existed, but there were no elements with which we would be able to proceed to a comparison of two equal or different sizes. The above "concepts" started to be expressed gradually in parallel with the development of the various events and we can suppose that there are still "infinite concepts" that have not been expressed and remain in silence. Anything that happens inside the Cosmos creates a series of new concepts. So, the concepts of force, straight or curved line, circle, speed etc., were manifested immediately after the formation of the first elementary particles. The concepts of the senses, the thought, the reasoning, etc., were manifested much later and after the development of the concept of life.

Space and Time "The First Components of Cosmos"

"Space" and "time" are two of the most basic "concepts" on which the foundation of the Universe, the Antiuniverse, the other Universes and Anti-universes and the whole Cosmos were built. Space and time preexisted as concepts in the "Cosmos" and even more they preexisted in huge quantities as nothing was needed for them to be created.

So, when we refer to the concept of "space" we mean a set of areas that begin from a certain position, extend towards all directions and end where a mental point can be led even by our thought. However, as our thought can be mentally led from any point in space to very big and endless distances, we reach the self-evident conclusion that space does not end but extends to vast distances and has infinite dimensions. Any point in space, "whenever it might be, either too far away or too close", becomes sequentially a new position from which a new space begins again, which in turn extends from this new position to infinity.

A lot of erroneous and unclear interpretations and theoretical suggestions have been formulated at times about the concept of space. However, we shall not analyze them as we should spend so much time for their analysis and rejection that we would be drawn away from the subject of our work. We shall simply accept the concept of space as a physical concept as it defined above, in the way that this concept is presented and becomes understandable, both simply in nature and in the manner in which a human brain can understand it.

Absolutely no process is needed for the formation of space. Space preexisted and is something that needs no energy in order to be formed. The various events take place inside space. For instance, matter, the stars, the voids between the stars, the Galaxies, the Universe, the Antiuniverse, probably other Universes and Antiuniverses, similar to or different from ours are formed and occupy increasingly larger areas of the existing absolutely void space, which was, remains and will remain continuously infinite.

The dimensions the events occupy inside space are huge but finite and very small compared to the total dimensions of the existing completely and absolutely void space. Thus, we could say that the size of a Solar system or a Galaxy or even a Universe, as vast as it might be, does not occupy even one thousandth of a thousandth in proportion of the total size of the existing unused absolutely void space.

The following peculiar case applies to the absolutely void space. Whereas the various physical dimensions start usually from zero and end at a certain real size or even to infinity, the dimensions of space start from the beginning of infinity and keep diminishing as vast areas occupied by events are deduced from it; however, these areas were, remain and will remain forever infinite.

Space has no limits. If we think of an imaginary boundary of space, as we can at least mentally go even further than that boundary, then it is automatically refuted. And if we assume that there is a real limit, as for example a certain condensed matter or a poisonous gas or something else that forms this boundary, as long as we can invent mechanical or other means that help us go beyond this boundary, then once more this boundary does not exist match more and is automatically refuted.

So this had to be said about space, which with time constituted the initial components of the "Cosmos". Now, about the concept of time we could say that "time" is the total of a series of successive moments within which the various events develop.

We divide the total time in three subsets, i.e., the "past", which is the time consisting of all the previous successive moments; the "present", which consists of just one moment that is continuously moving from the past to the future; and the "future", which is the total time that all the posterior moments will form.

In what concerns the first moment in the past, if we use the theoretical mathematic concept of "retroactivity" and assume that for every moment of the past there was always a previous one, we reach the simple and at the same time very basic conclusion that the first moment in the past never existed or, if it existed, it has to be lost back in the infinity of moments of the past.

So, as the first moment in the past is lost back in the infinite, we reach the conclusion that the moments of the past are infinite too and, therefore, the total of the past is also infinite. We are not going to mention specific measures about the total of the past time here, because this issue is not part of the subject of the present work. However, we have to make clear that we did not accept the concept of infinity for the past just in order to define something very big, but we accepted it according to its purely theoretical mathematical meaning in order to define something endless or something "infinite".

Commenting the above paragraph and in order to show even approximately how we perceive the infinite size of past time, we consider that there is no size that could express the totality of past time. Millions, billions, trillions of years are very small numbers to approach even the least of past time. For instance, if we suppose that in every single year in the Universe just one atom was formed, even then the total of atoms of our Universe is not enough to express the total of past years.

At this point, we should also clarify the big difference between the concept of the first moment of the past and that of the first moment of the beginning of the creation of our Universe or another event in the "Cosmos". These concepts, that is the "beginning of time" and the "beginning of the creation of a Universe" or an event, are two completely different notions.

More specifically, the concept of the beginning of time is a theoretical mathematical concept the beginning of which is lost in the infinity of the past time, whereas the concept of the beginning of the creation of our Universe or another event is a real concept the beginning of which, as remote as it can be, remains a physical event, which started at a certain, specific and finite moment in the past.

The present, as mentioned in the definition, consists of only one moment "the present moment", which is modified and displaced continuously from the past to the future. It is the moment that separates and at the same time connects the past and the future; it is modified so rapidly that until we realize that a moment of the future belongs to the present it has already become past.

Now, in what concerns the future time, if we proceed to the same rationale as we did for the past, we can once more conclude that the total time of the future will also have infinite theoretical dimensions and that the end of the future is lost too, into the infinite of the posterior moments.

Therefore, the future, too, as a time set, is a theoretical mathematical notion for which there is no end. A certain event can end at a certain moment, but the moments of the future will not end with this event, as we supposed for the past too, a posterior moment will always correspond at each moment of the future and will be related to certain new events. For instance, if we take the extreme case of the destruction of our Universe, even then, time will not end but will continue its course with the evolution of the other Universes and Anti-universes.

The basic difference between the concepts of past and future is that, whereas we have many data about the events of the past and we can easily and safely study them, we have absolutely no data about the events of the future. The only thing we could suppose with some degree of certainty is that the end of time in the future, as mentioned above, should be lost in the infinity of the future moments.

In simple human reasoning, the notion of time is instinctively understandable as a set of moments which we realize due to various events and we identify them with those events. However, in the present study, we shall distinguish the notion of events, which is real and different from the notion of time, which is purely theoretical, and we shall accept that time can exist without events, whereas, on the contrary, events cannot exist without time.

At this point, I would like to emphasize, as I have already analyzed, the huge difference between the beginning of time and the beginning of the creation of our Universe. These two notions are completely different. The beginning of time is a theoretical concept lost in the infinity of the past, whereas, on the contrary, the beginning of the creation of our or any other Universe is a real and finite notion that corresponds to a particular point in time, even though this point is several billion years back from the present.

Many researchers make the mistake of correlating the beginning of time in the past with the beginning of the creation of our Universe and the eventual end of our Universe, "if there will be such a case" with the end of time. In this case they make a very big mistake correlating two completely different concepts, that is a theoretical concept, "the beginning or the end of time" with a real concept, "the beginning or the end of a Universe or any other event". This mistake is made because the researchers identify time with events. The result of this correlation is that a lot of peculiar, unclear and incomprehensible interpretations, ideas and theories are written about time.

But, I believe I have tired you enough with all this talking and these generalities about the Universe, the Antiuniverse, the Universes, the Antiuniverses, the Cosmos and the concepts of space and time. For this reason, I shall end this chapter letting a "Cosmos" being constituted exclusively by the infinite "time", an infinite absolutely void "space" in a latent condition and the infinite obscure "notions" as described above. All those shaped the conditions that existed at the exact moment when the first events in the Cosmos started, which, according to the "theory of the chain reaction" were the formation of the first elementary particles.

But before I continue with the presentation of the "theory of the chain reaction" and the description of the first events of the beginning of the creation of the Cosmos and our Universe, I shall insert a parenthesis exposing the chapters about "The evolution of human knowledge about the Universe", "The Universe as it is known today", "Matter, elementary particles and fundamental forces" and "The cosmological theories and the theory of the big bang".

This parenthesis for the presentation of the four chapters above will update us about the development of human knowledge concerning the creation of our Universe till our days. At the same time it will help us understand more efficiently the "theory of the chain reaction", the understanding of which will finally provide us a more complete image of the Cosmos.

Chapter 2

THE EVOLUTION OF THE HUMAN KNOWLEDGE ABOUT THE UNIVERSE

In 350 B.C., Aristotle maintained that the Earth is spherical, that its center is identical to that of the Universe and that the Earth does not move but everything moves around it. About twenty –20– centuries passed and it was in 1514 when a Polish priest, Nicolas Copernicus, suggested that the Sun does not move and that the other stars move around it. Today it is believed that the Sun too moves relatively to a certain point in our Galaxy and that our Galaxy, in turn, moves relatively to another point in our Universe.

THE KNOWLEDGE OF THE PRIMITIVE PEOPLE ABOUT THE UNIVERSE

Man's efforts to interpret the Universe and the events that take place around him, started from the very moment humans started to have an even primitive, rudimentary rational thinking.

For primitive humans, the Universe consisted of several large flat pieces of land which they later named "plains" and several bigger or smaller protrusions which they named "mountains". The plains and mountains were in turn surrounded by large quantities of water, the "sea", which at that time nobody could understand where it started and where it ended.

All the above were covered by a large cover which was at times bright and at other times dark. Sometimes, when the cover was bright, it was light blue and sometimes it was gray. But many times a larger light appeared too.

Later and in order to distinguish them, humans named as we have already mentioned, the flat surfaces plains, the protrusions mountains and the masses of water "sea". They named the surface that covered the plains, the mountains and the sea "sky", the period during which the sky was bright as "day" and the period it was dark as "night". Also, they named the tiny lights they saw in the sky at night as "stars" and the two bigger lights that shone during the day and the night as "Sun" and "Moon", respectively.

However, humans wondered and could not understand what the stars were. What were the Sun and the Moon? Why was the sky at times bright or dark, blue or gray? They could also not understand how all those plains, the mountains, the sea, etc., were formed.

In this way the first human experiences, the questions, as well as the interpretations about the Universe started to form gradually and in general. The interpretations they gave about what happened around them were primitive and totally insufficient, but they were correct, because they were based on rational reasoning and direct observation.

As humans did not dispose of the necessary infrastructure, the instruments required or advanced reasoning, they were not able to study the various events thoroughly and therefore they were restricted in examining, judging and trying to interpret what really happened around them with simple reasoning and direct observation. This is also the reason for which we can maintain that what humans knew at that time was quite insufficient, but very accurate.

THE MYTHOLOGICAL PERCEPTIONS AND THE RELIGIOUS INTERPRETATIONS

Then, as human reasoning improved, those simple interpretations were increasingly supplemented. Instruments were manufactured which measured the various physical dimensions, interpretations were given for several physical phenomena and gradually we reached the current level of knowledge about the Universe and the Cosmos in general.

Let us, however, see how the initial knowledge and experience evolved in parallel to progress so that from a simple image about the Cosmos and the Universe humans reached the current levels of evolution, knowledge and experience.

We mentioned above that humans observed what happened around them but were not able to understand why that happened. For instance, they saw the stars but did not conclude anything from them. So and as a first step towards an explanation of what happened, all people developed mythology.

Each people had their own myths about who created the Universe, what the stars are and where they come from, what the various physical phenomena mean, etc. Certainly, the people who tried to organize their lives and developed civilization through hunting, agriculture, cattle-breeding, the sea and commerce were pioneers in this effort. Among those people were the Persians, the Babylonians, the Chinese, the Indians, the Egyptians and the Phoenicians.

So, for example, the Phoenicians and the Babylonians believed that their God fought with a dragon and their God was the winner, tearing the dragon in two pieces and therefore creating the sky and the earth. Then, he placed the stars in the sky, the Sun and the Moon and thus he created the Universe.

The Indians and the Egyptians believed that in the beginning there was nothing but chaos. Darkness prevailed everywhere and strong wind blew. Matter was created from the wind and the dark and then the Universe, the Cosmos and Life were created.

The northern people, the Norwegians and the Danish, believed that in the beginning there was nothing but the chaos, which manifested through cold, fog and fire. Cold and fire joined and created their God, Ymir. A cow was born from fog; Ymir was nourished with its milk. Then other gods were also born and separated land from water and fire. So they created the Sun and the Earth and, gradually, they formed the Universe.

The Indians found a peculiar and very simplistic way to answer the question of how the Cosmos was formed. According to the Indians, in the beginning only the Chief God, Tiravat existed, who, after creating the Cosmos, created the other gods and gave them various positions in the sky in order to govern the Cosmos. He made some of the gods he had created to support the sky he had already formed.

At the same time as mythologies explained, in their own way, the creation of the Universe and the cosmological evolution, the religious feeling also developed in peoples. As humans were not able to react and explain adequately the various physical phenomena, they felt the need to be protected by a certain "Upper Force". So, the "concept" of God was gradually created, representing what they themselves were not able to achieve. Then, in a broadening of the concept of God, the religions were born, which at first played the most important role in Cosmological issues.

All religions, without any exception, believe that everything started from their God, who is always the highest force protecting people and being by their side at all difficult moments as he knows and dominates everything and is the one who created and guides the Cosmos and the Universe.

The religions of the ancient northern people, who were identified with mythology, accepted something similar to their mythology, i.e., that, at the beginning, there was nothing inside the Cosmos, neither sky nor land nor sea; chaos was everywhere, represented by cold, fire and fog. The joining of those elements as a first creation was their God, who, using fog as a raw material made a cow in order to provide for his food with her milk. Then, he made the sky, the land and the water, he separated the land from the water and he made trees and animals and thus, gradually, he created the Universe.

THE ANCIENT GREEK PHILOSOPHERS AND THE FIRST SCIENTIFIC EXPLANATIONS

The first attempts for a scientific explanation about the Universe were made by ancient Greek philosophers who, after having studied the existing knowledge, tried to provide some reasonable, complete and organized explanations about what happened around them.

In 600 B.C. the Pythagoreans were the first to study the Universe, which they named "Cosmos", meaning ornament. In this way they wished to express their admiration for the beauty and harmony that prevailed around them. They believed that in the beginning there was the chaos and the Creative God made everything. The Pythagoreans were the first to find out that the Earth is round. Until the Pythagorean era, people believed the Earth was flat.

Then Aristotle, who is considered to be the founder of sciences, left around 350 B.C. a vast work in which the treatise about sky is distinguished. In his work, Aristotle proved that the Earth is really round and generalized this conclusion for all celestial bodies.

The smart way and the very simple arguments he used to prove the round shape of the Earth are exquisite. Characteristically, he used three arguments. The first of these arguments was based on the phenomenon of the moon eclipses. Aristotle said that the eclipses are due to the interposition of the Earth between the Sun and the Moon, exactly as it really happens. If Earth was flat and not a sphere, its shadow on the Moon would be an ellipse and not a circle as it is in reality. A second argument was that from a remote position in the sea we first see the mast of a ship and then its hull; this is again due to the round shape of the Earth. But he had a third argument also about the round shape of the Earth. This argument was based on the Polestar or Lodestar, which, due to the round shape of the Earth, was seen lower on the celestial body compared to regions located more southern. In particular, he mentions the difference in height of the Lodestar between Greece and Egypt.

In 300 B.C., Aristarchus of Samos, who is considered one of the founders of astronomy, claimed that the Earth moves around the Sun and around itself at the same time; however, due to the lack of instruments at that time he was not able to formulate the appropriate arguments or proceed to adequate experiments by which he would be able to found his opinion.

On the contrary, those who supported the view that the Earth remains still –as we describe analytically in the next third section of this chapter– said that if the Earth moved we should be feeling a continuous current of air that would be produced and at the same time we would feel the motion of the ground. They also maintained that the idea of a moving Earth was not consistent with the perceptions of the time about gravity. Another argument supporting the view that the earth remained still was the apparent immobility of the stars. If the earth covered those distances around the Sun, which at that time were huge, we should also have respective motion of the stars.

Thus, Aristarchus' idea, even though correct, was not accepted and about twenty centuries passed before Copernicus supported it again as we describe later. Aristarchus calculated also the distance between the Sun and the Earth, but due to the lack of appropriate instruments, his calculation, though theoretically absolutely correct, was not so precise and therefore the distance he calculated was much different from the real distance.

Then, from 300 to 150 B.C., many Astronomers, Mathematicians, Physicists and Scholars –among who Eratosthenes' brilliant personality was distinguished– worked with simple ways and calculations, as described in the next section in order to found our fist ideas about the Universe; ideas that influence our theoretical knowledge even today.

Around 150 B.C., Hipparchus of Nicaea of Bithynia opened new horizons to astronomy and is considered today as the founder of Mathematical Astronomy and Geography. He invented many astrological instruments, some of which are used even today. Hipparchus wrote seventeen works in total, but unfortunately only two works survived. In these works the calculation of the duration of the Solar year is included, which he calculated with surprising precision, finding 365.246 days instead of 365.242 that is the current closer approximation!

Note: The fact that the establishment of leap years every four years and of one additional leap year every one hundred and twenty five years is due to the decimal part, 0.242 of the Solar year and should be taken into account.

The calculation of the length of the biggest circle of the Earth is also included in Hipparchus' books; he found that it was 252,000 stadiums or 39,960 kilometers, very precise compared to the current calculation of 40,000 kilometers! These calculations could be considered "extraordinary calculations" as they were done with the use of the instruments of that time.

One of the surviving Hipparchus' works is a list of about 1,100 stars with a lot of information about their position and their brightness. Hipparchus describes also in this list the constellations in which each of those stars belonged. This list was the second after the list made by the Chinese, which included about 800 stars.

How Ancient Greeks Calculated the First Basic Measurements of Our Solar System

In the previous section we saw that Earth's round shape was one of the first scientific ideas founded by the Ancient Greeks. After the establishment of the idea that the Earth is round, the route for the calculation of its size opened, i.e., for the calculation of the length of its maximum circle, its diameter and its radius. Next, these calculations helped in finding the distance between the Earth and the Moon, the size of the Moon, the distance between the Earth and the Sun and the size of the Sun.

In the next paragraphs we are going to provide a very brief description of these calculations in order to see how the Ancient Greeks made the above measurements with very simple observations, simple calculations and rudimentary instruments and how they established the first dimensions of our Solar system, which constituted also our first scientific experiences and measurements of the Universe.

So, after founding and proving the view, that the Earth is round, the Ancient Greeks noted that in Syene, a town of Egypt, located south of Alexandria, on June 21 at noon, every year, the Sun was reflected on the water of all the wells of the town. This meant that, at that time, the Sun was overhead Syene and that its rays fell perpendicular to the surface of the Earth and passed from its center, as shown in Figure 1.

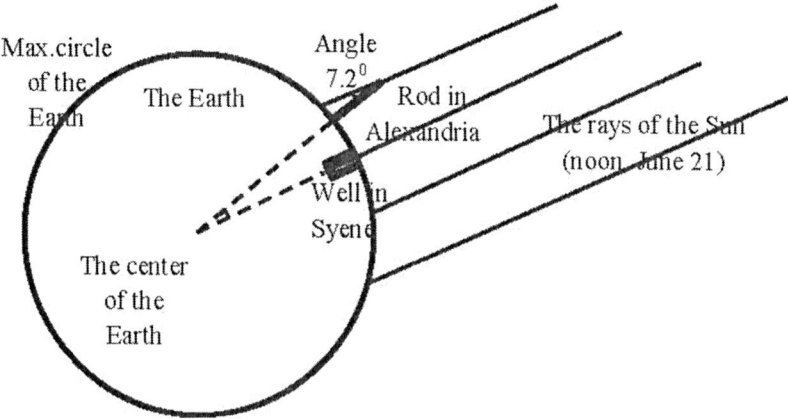

Figure 1. From the calculation of the maximum circle, the radius and the diameter, of the Earth.

Taking advantage of this fact and using a small rod which he placed vertically, Eratosthenes of Alexandria measured the angle formed by the rays of the Sun and the rod at that moment –that means, the moment when the Sun was overhead of Syene–, in Alexandria. He determined easily that this angle in Alexandria was 7.2°.

Using this measurement together with the geometrical property of parallel lines about the equality of the alternate angles, as shown in Figure 1, Eratosthenes found that the angle formed by the rays of the Sun and the rod was equal to that formed by "Alexandria", the "center of the Earth" and "Syene", of 7.2°. After this observation, the calculation of the maximum circle, the diameter and the radius of the Earth, was very simple and could be understood even by someone who did possess even elementary knowledge of geometry.

This means that Eratosthenes, knowing that the distance between Syene and Alexandria was 5,000 stadiums and that this distance on a max. circle of the Earth corresponded to an angle of 7.2° as he had measured it above, calculated that to an angle of 1° a distance of 5,000/7.2° = 694.4 stadiums would correspond and therefore the size of the maximum circle of the Earth, which corresponds to an angle of 360° is 694.4 X 360 = 250,000 stadiums or 46,000 kilometers. More precise current calculations showed that the size of the maximum circle of the Earth is 40,100 kilometers.

Then, the calculation of the radius and the diameter of the Earth were very simple with the use of our well known formulae: radius of the Earth = maximum circle of the Earth/2X3.14 = 6,340 kilometers and diameter of the Earth = radius of the Earth X 2 = 12,680 kilometers.

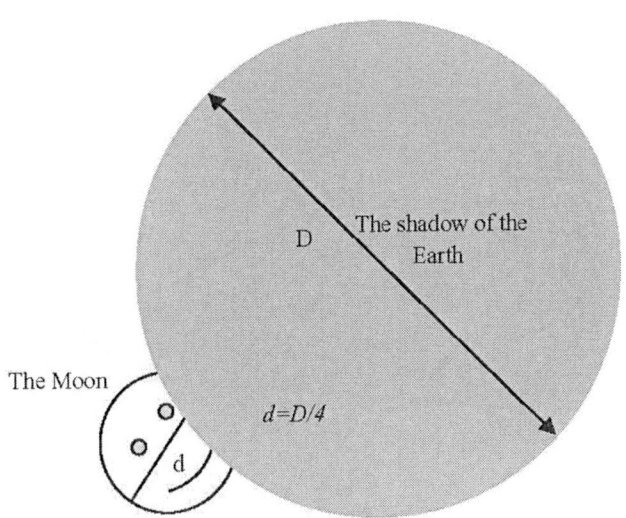

Figure 2. The positions and the relative dimensions of the Moon and the shadow of the Earth during the eclipse of the Moon.

After the successful calculation of the dimensions of the Earth, the route to the calculation of the dimensions of the Moon opened when the Ancient Greeks noted that, during the eclipses of the Moon, the shadow of the Earth that fell on the Moon and caused the eclipse was four times the size of the Moon, as shown in Figure 2. So they thought that the dimensions of the Earth should also be four times those of the Moon.

So, taking this observation into account, they easily calculated that the maximum circle of the Moon should be 46,000 kilometers/4, i.e., 11,500 kilometers, its radius 6,340 kilometers/4, i.e., 1,585 kilometers and its diameter 1,585 X 2 = 3,170 kilometers.

More precise measurements that were performed later with better instruments yielded the following results of the dimensions of the Moon: circumference of the maximum circle 10,940 kilometers, diameter 3,480 kilometers and radius 1,740 kilometers.

Although the above calculations were fairly simple, the calculation of the distance between the Earth and the Moon held the record of simplicity. I suggest to the reader that we proceed to this measurement and the relevant calculation together, while reading this book!

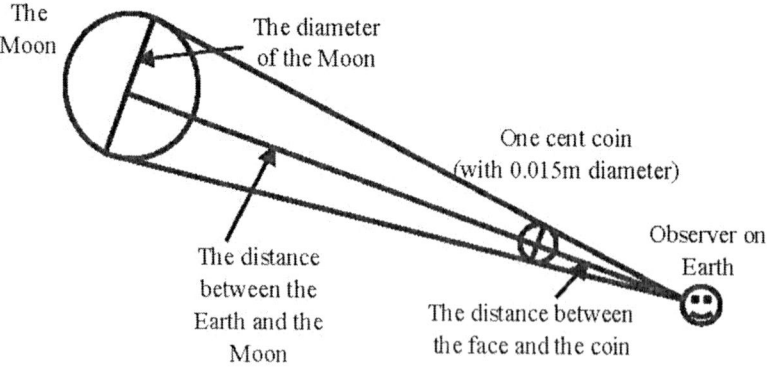

Figure 3. Figure depicting the calculation of the distance between the Earth and the Moon.

There is no need to read the above paragraph again, because as you have correctly understood, my suggestion is to proceed to the measurement of the distance between the Earth and the Moon together now, at this very moment!... provided that we possess a coin of one cent –which you will see has a diameter of 1.5 centimeters or 0.015 meters if you measure it– and are at a certain place, from where we can see the Moon.

Starting the measurement of the distance between the Earth and the Moon, try to place the one cent coin you have in your hand at such a distance from your face that it will cover the whole disc of the Moon, as shown in Figure 3 above. Then, measure the distance between the coin and your face. If you find a distance of approximately 1.5 meters you are correct and the measurement of the distance between the Earth and the Moon has finished. Yes, you understood correctly, the measurement has already finished. From the formula distance between the Earth and the Moon = diameter of the Moon/diameter of the coin X distance between the coin and the face (formula that applies according to the use of the geometry rule about the proportionality of the sides of similar triangles, as shown in Figure 3), with the use of the dimensions we have already calculated we have: distance between the Earth and the Moon = 3,480/0.015X1.5, i.e., 348,000 kilometers. More precise measurements yielded a result of 384,000 kilometers for the distance between the Earth and the Moon.

Of course, not all calculations, as we shall see next, were so simple. However, we can say that in this case science helps us both in our simple and in our difficult problems. So, after the calculation of the distance between the Earth and the Moon, the calculation of the distance between the Earth and the Sun was much more difficult. However, thanks to a brilliant idea of Anaxagoras and Aristarchus, the Ancient Greeks calculated the distance between the Earth and the Sun as follows: when we have a half-moon (that is, when from the Earth it seems that the Sun lightens only half of the Moon), the Sun, the Moon and the Earth must form a right triangle, exactly as shown in Figure 4, below.

Therefore, in this case, if we measure the angle formed between the lines connecting the Earth and the Moon and the Earth and the Sun and as we have already measured the distance between the Earth and the Moon, we can easily draw the right triangle Sun-Moon-Earth (or use trigonometry) and calculate the side Earth-Sun of the triangle, which will also be the corresponding distance. Aristarchus measured the above angle and found it to be 87°; then, he drew the right triangle Earth-Moon-Sun and calculated that the side Earth-Sun of the triangle he drew is 20 times bigger than the respective side Earth-Moon, which means that the respective distance between the Earth and the Sun is also 20 times bigger than the distance between the Earth and the Moon;

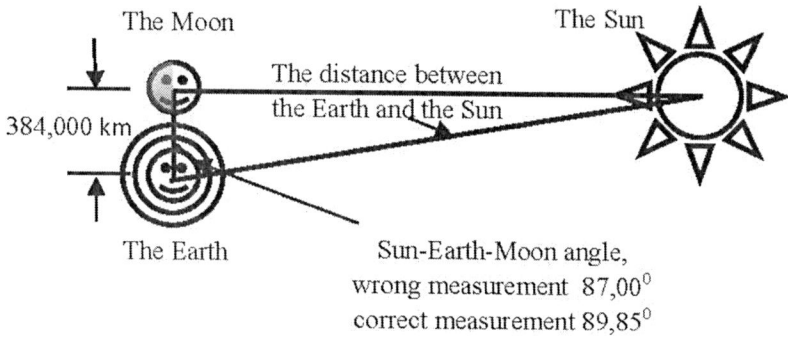

Figure 4. Figure depicting the measurement of the distance between the Earth and the Sun.

However, we have already calculated this distance and so he calculated then that the distance between the Earth and the Sun = the distance between the Earth and the Moon X 20 = 320,000 X 20 = 6,400,000 kilometers.

In fact, the angle Sun-Earth-Moon at half-moon is 89.85° and not 87°, as Aristarchus had measured, and therefore the distance between the Earth and the Sun is 400 and not only 20 times bigger than that of the distance between the Earth and the Moon, as believed. In this case, we see that the result of Aristarchus' measurements was very different from reality, whereas the method he used for the measurement of the distance was correct, very brilliant and valid till our days. We shall discuss the case of this quite serious experimental mistake in the measurement of the above angle in the next section. However, more precise measurements made with the use of the exactly the same method defined the distance between the Earth and the Sun as 150,000,000 kilometers.

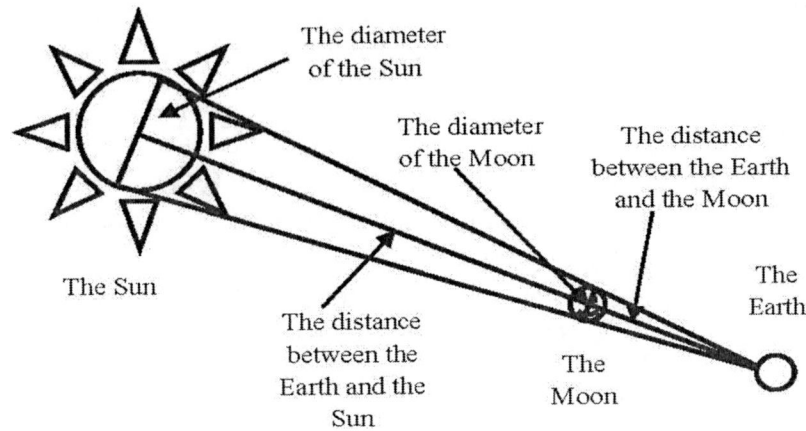

Figure 5. From the calculation of the dimensions of the Sun, based on a total eclipse.

After the measurement of the distance between the Earth and the Sun, it was fairly easy to calculate the dimensions of the Sun. In this case, the Ancient Greeks used exactly the same method with which they measured the distance between the Earth and the Moon, as we described above, with the help of the one cent coin, etc. The respective calculation is made by putting the diameter of the Sun in the same formula instead of the distance between the Earth and the Moon.

However, Ancient Greeks kept investigating things scientifically and they were not satisfied with the above method not only for the measurement of the dimensions of the Sun, but in order to avoid multiple measurements they thought of another similar but simpler method. They noted that in a total Solar Eclipse, the Moon fits exactly on the disc of the Sun. They calculated thus the dimensions of the Sun in exactly the same way they had calculated the distance between the Earth and the Moon, except that in this case the Moon itself played the role of the one cent coin, as they already knew its distance from the Earth.

Applying again, as shown in Figure 5, the formula based on the rule of proportionality of the sides of similar triangles, we have: Sun diameter =

distance between the Earth and the Sun/distance between the Earth and the Moon X Moon diameter. Then, replacing in the above equation the data we have already calculated we have: for the Sun diameter = 150,000,000/384,000 X 3,480, i.e., 1,360,000 kilometers. More precise measurements yielded a result of 1,390,000 kilometers for the diameter of the Sun.

The above calculations and the methods used, established the first scientific ways for the measurement of the first dimensions of our solar system; dimensions that also constituted our first experience of the greatness of the Universe.

Closing this section, we can express as a final conclusion, our admiration to the Ancient Greeks who, using the above simple ways, with rudimentary instruments and simple reasoning, opened the way for the scientific study of the creation and the function of the Cosmos.

THE SCIENTIFIC THINKING OF THE ANCIENT GREEKS

The above extraordinary scientific achievements of the Ancient Greeks in the measurement of the dimensions of our Solar system reflect the great progress of scientific thinking in Ancient Greece. This progress was based on the simplicity, the understanding, the reasoning, the philosophy, the mathematics, the measurements, the observation and experimentation.

Of course, long before Ancient Greeks, other people, such as the Babylonians and the Egyptians, had made many great observations about the sky, but their observations did not go any further from that. So, in this case, the measurements and the collection of data about hundreds of stars kept repeating the same story and the same trivial quality that classifies all those observations in the class of technological observations and not in that of scientific achievements. Concerning those observations, we could say that science, in order to advance, needs observations just as a house needs materials in order to be built. However, if there is nobody to arrange those materials, the result will be a pile of sand, cement, bricks, iron rods, etc.

So, we could say that the progress that had been made until the apparition of the Ancient Greeks was mostly technological. Just as the Babylonians had made many measurements of the sky, the Egyptians in their turn, many years before the Ancient Greeks, had built the Pyramids and had invented many instruments for measurements and had constructed many other products and tools for everyday life. However, all the above were just technological advances. But technology by itself could not lead to the appropriate progress. Something that would establish those technological developments was needed and this was the scientific progress that would found and establish the technological advance. So the Ancient Greeks could boast that they established this scientific development.

THE END OF THE ANTIQUITY AND THE ESTABLISHMENT OF THE GEOCENTRICAL SYSTEM

This period ended in 180 A.D. by Claudius Ptolemy who studied and improved the model of the Universe given by Aristotle about five hundred years prior. Ptolemy reviewed Aristotle's ideas and provided the astronomical model of the Geocentrical system as shown in Figure 6 on the next page with the Earth steady at the center of the Universe and the Sun, the Moon and the planets and the other stars moving around the Earth.

As mentioned in the second section of this chapter, the reason for which Ptolemy's Geocentrical system was established and the Heliocentric system suggested by Aristarchus was rejected was the apparent failure of the Heliocentric system to stand up to a detailed scientific review with the tools of that time. The Heliocentric system may be the correct one or even be closer to reality, but at that time it could not be proven precise.

Indeed, people on Earth feel they are not moving and that the ground on which they stand remains steady. At the same time, the steadiness of the ground can be established by the fact that there is no continuous wind suggesting a certain movement in the environment. So, on what grounds could people at that time accept that the Earth is moving? Of course, this idea that the Earth is still is wrong. In reality the Earth moves and at the same time the atmosphere and the ground and all the objects move too so that we are not aware of this motion.

Another indication that the Earth did not move was that a motion of the Earth was not compatible to the understanding of that time about gravity. At that time, they noted that all bodies were attracted towards the center of the Earth. But, as it is natural that bodies are also attracted to the center of the Universe, the Earth would be the center of the Universe. If the Sun were the center of the Universe, then all bodies should be attracted to the Sun. Therefore, a free object on Earth would not move towards the center of the Earth, but towards the Sun.

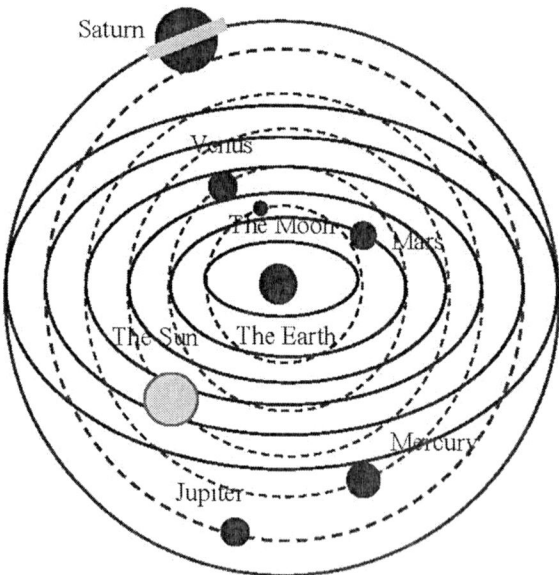

Figure 6. Ptolemy's Geocentrically System.

This rationale was certainly wrong as the bodies are not attracted only to the center of the Earth, but as we know, they are also attracted among them proportionally to their size and distance. Thus, a body is attracted both to the center of the Earth and to the Sun. However, in the case that the body is close to Earth, the attraction to the Earth prevails due to the short distance and it seems like it is attracted only from the Earth. So, a body close to the Earth is attracted by it because it is close to it and not because the Earth is the center of the Universe. A body close to the Sun will be attracted more by the Sun.

A third reason for considering the Earth still was the apparent zero displacement of the various stars. If the Earth covered long distances around the Sun, we should see the Universe differently during the year, because we would be watching it from different positions.

This means that along with the motion of the Earth the stars in the sky should also move. However, with the instruments of that time, the stars seemed still and therefore the Earth should be still too. In reality, however, it is not correct that the stars remain still, as we described in the third section of this chapter that all stars move but seem still due to their very long distances from the Earth. So, as we see that all stars move, if we watch more attentively, it is natural to understand that the Earth moves too.

Certainly, the model of Ptolemy's geocentrically system was wrong. However, it prevailed in humanity for a very long period of about fifteen centuries. This happened because the geocentrically system was apparently correct and helped the people measure distances accurately and orientate themselves easily, mainly at sea.

The most important reason for the prevalence of Ptolemy's geocentrically system for so many centuries was the fact that it resolves infallibly all the issues about the Terrestrial environment, without any exceptions. So, for many centuries no mistakes were noted that would impose a modification of the system.

THE EVOLUTION IN THE KNOWLEDGE ABOUT THE UNIVERSE IN THE MIDDLE AGES

The era after the fall of the Roman Empire –about 400 A.D.– until the dawn of the era of the Renaissance –about 1450 A.D.– was named the

Middle Ages –or Dark Ages– and it is the period between the Antiquity and the Renaissance.

During the Middle Ages there was no progress in sciences in Europe and therefore nobody expected any specific development in physics, cosmology or astronomy.

Those who continued the research and the advance that had been achieved during the ancient era were the Indians, the Chinese and the Arabs. During the Middle Ages, "Astronomy" or "the science of the sky" as they called it advanced fairly well among these peoples. Observatories were constructed, equipped with perfect instruments that measured with fairly good precision, maps of the Stars in the sky were drawn and the dimensions of the Earth were calculated with precision.

Now, in parallel with astronomy, "Astrology" was also developed, which tried to connect the movements and the positions of the stars with various predictions. And as several of the astrologers' predictions happened to prove often true, astrology started to strengthen more and more.

So, the astrologers connected the appearance of several bright stars with good or bad things that would occur on Earth, with the birth or death of kings, with wars and various diseases or disasters. It is said that the three wise men (Magi) with the gifts were astrologers of that time and that the appearance of a very bright star, a comet perhaps, led them to the search of a newborn king.

The interest of the people of that time to learn something about the future led them to extended and desperate observations of the stars in the sky. So, regardless of their wrong or correct prophecies, the observations of the motion and the position of the stars in the sky were proven valuable for the advancement of Astronomy and Cosmology, exactly as it happened with the observations of the alchemists, which were proven valuable in Chemistry.

During the period of about 1100 years from Ptolemy's era to the Renaissance, the perception that the Earth was still and firm at the center of the Universe prevailed in Europe. The known celestial bodies were moving around it; they had been formed only to lighten it.

The various observations and the formulation of ideas were made solely by monks with a certain level of education, who knew how to calculate several elements of time, the seasons, the rise and the setting of the Sun, etc. There was an established status of knowledge that did not accept any changes at all. If someone expressed a different idea they were persecuted by this established situation which in this way managed to impose its erroneous ideas and keep the advancement and the development at the levels of Ptolemy Claudius' era.

THE EVOLUTION FROM THE TIME OF THE RENAISSANCE UNTIL NOW

About 1500 A.D., Nicolas Copernicus –1473-1543–, monk and astronomer, after 36 years of work and observation, studies and calculations on the issue, formulated the theory of the Heliocentric model for our Solar system, with the satellites moving in circular orbits around their planets and then the planets, together with their satellites, moving in circular orbits too around the Sun; a movement performed also by the Earth, which until then was considered still at the center of the Universe.

Copernicus' Heliocentric theory was the first blow to Ptolemy's Geocentrically Theory. The second blow came from the Italian physicist and astronomer Galileo –1564-1642–, who perfected the telescope and managed, with its use, to observe in detail the rotation of the Moon around the Earth and study the movement of several other satellites that rotate around their planets. Then, he proved that these planets with their satellites rotate, in their turn, around the Sun.

However, the established situation of that time was by no means willing to accept Copernicus' and Galileo's new ideas about the Universe, although they are ideas that today can be understood by every simple mind. All those who believed these ideas at that time were fiercely persecuted. And there are many cases of great scientists who were violently forced to change their ideas, whereas those who did not accept to do it suffered unheard of tortures.

Of course we all know Galileo's case, when the disclosure of his views and ideas in his work "Dialogue" caused his arrest in July 1633; under the threat of tortures he was forced to apologize on his knees in court. It is said that at the moment Galileo declared in tears his repentance with his hand on the holy bible, as the judges had requested, he spelled the well known phrase "and yet it moves" meaning the motion of the Earth around the Sun. The judges then condemned Galileo to imprisonment, but due to his age, they lessened his sentence to house arrest, where Galileo continued his research even after his conviction.

However, despite the severe persecutions against Copernicus' and Galileo's ideas, research continued. Astronomy reached a level at which, with the help of the German astronomer Kepler –1571-1630–, it could describe even the laws according to which the planets move around the Sun.

At that time it was also found out that the orbits of the planets around the Sun and the satellites around their planets were not exactly circular, as it was believed, but elliptical. However, the established situation of that time was by no means willing to accept anything of all these advances and remained loyal to its erroneous views.

In 1667, Newton –1642-1727– formulated the international law of gravitation which says that: "Material bodies are attracted with a force proportional to their masses and inversely proportional to the square of the distance between them", which means that the bigger two bodies are the greater will be the attraction developing between them.

With the formulation of the international law of gravitation by Newton, the theoretical explanation about the movement of the planets around the Sun was finally given and at the same time the phenomenon that they remain at their positions in the sky without colliding was explained. Thus, the established situation of the time could no longer disclaim these facts and that resulted in the gradual establishment of the new ideas that opened new horizons in research for the explanation of the phenomena in the megacosm.

The Evolution of the Human Knowledge about the Universe 29

So, gradually, the Heliocentric System was established, according to which the satellites move in elliptical orbits around their planets and the planets, in turn, move together with their satellites in elliptical orbits around the Sun, as shown in Figure 7 below.

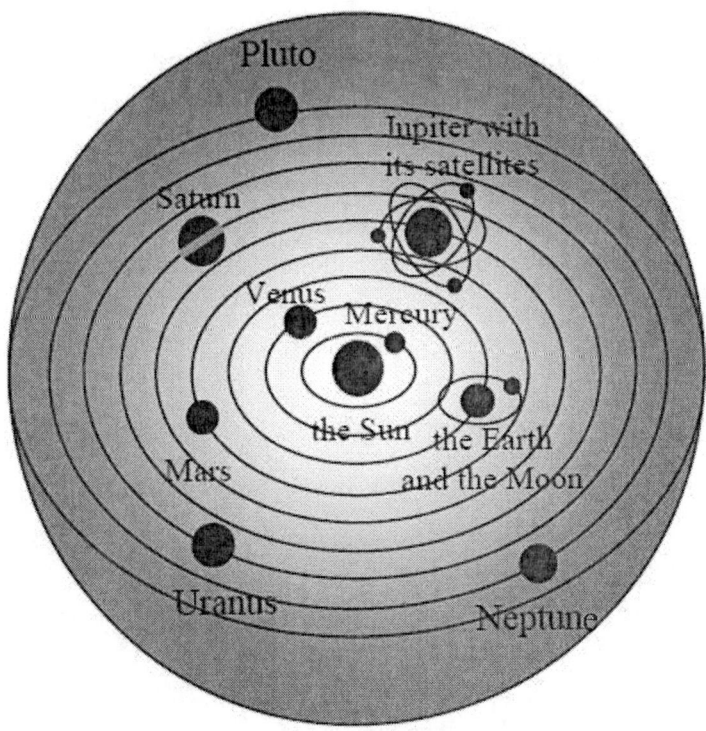

Figure 7. The Heliocentric system with the planets and the satellites of the planets.

Then, as the investigation of space progressed, it was found that the Sun moves inside our Galaxy and that the Galaxies move and there are drawn away from an imaginary center of our Universe, which is the point from where its creation started according to the theory of the "big bang", as it is described in chapter five.

All the above achievements concerned the study and the investigation of the "megacosm", that means the world of the large, but closed distances. These distances started from the dimensions of the molecules and the atoms and from the distances between them and came up to the distances between the satellites and the planets in our Solar system or up to the distances between the Stars within our Galaxy. Then, the theoretical investigation of the "World of infinity" started. In this world the distances are now extending further than the distances between the stars inside the Solar system and the distances between Galaxies inside the Universe.

In parallel to the research about the world of the infinity, another race started, concerning the study of the microcosm, i.e., the world of the very small dimensions, which are inside the atom and inside the nucleus of the atom.

So, gradually, a series of fast advances started in the investigation of the "Cosmos of the infinity" and the investigation of the "microcosm", which are still evolving, as we describe analytically in the two sections that follow.

THE EVOLUTION OF THE KNOWLEDGE ABOUT THE COSMOS OF THE INFINITE

During the first decades of the past century the image of the Universe was once more reversed. Until then the Universe was considered static; this means that it was thought to be an eternal and unmodified Universe with the Stars and the Galaxies occupying fixed positions in it. And the cosmologists thought that this image of a constant and unmodified Universe would last forever.

However, the image changed when in 1929 an astronomer named Hubble, based on the "Doppler" phenomenon, according to which "the light emitted by a moving object changes its form depending on the velocity of the object", made a revolutionary discovery about the Universe. He observed that the Galaxies draw away from each other. The speed of their drawing away is proportional to their distance, which means that, besides being drawn away from each other they are also at the same time drawing away from an imaginary center, the center of the Universe. This center, as we shall see later, is considered identical to the point from which the big explosion began that started the creation according to the homonymous theory of the "big bang".

So we live in a Universe that keeps expanding. This expanding Universe is like a big balloon that is continuously getting inflated. The Galaxies are small shapes on the surface of the balloon. As the balloon gets more inflated, the shapes on its surface get bigger and bigger and drawn away from each other, proportionally to their distance. This is exactly what happens with the Galaxies in the Universe.

Based on the Doppler phenomenon, Hubble was able to measure even the speeds by which several Galaxies are drawing away. In this way it was measured that the Galaxy of Virgo, which is relatively close to us, is drawing away at a speed of 11,200 Km/sec, whereas the Galaxy of Hydra, which is at a bigger distance, is drawing away at a greater speed of 61,000 Km/sec. These measurements show us also that the remote galaxies are drawn away at speeds that approximate the speed of light.

So it is indisputable that the Galaxies are drawn away both from each other and from an imaginary central point of the Universe. These led scientists to the conclusion that initially the whole Universe was concentrated in only one spot from where a big expansion started at a very high speed – so high that it looked like an explosion. It is believed that the Universe was formed from this explosion.

There is no doubt that Hubble's absolutely correct observations led us to the idea of the big explosion. However, the theory of the "big bang", as it is usually called, although answering the issue of the movement of the Galaxies, does not provide answers for many other queries. The queries that remain unanswered by the theory of the "big bang", some of which I discuss in chapter five, led scientists to the search of a new theory that would be based on Hubble's observations about the movement of the Galaxies but would at the same time answer more convincingly and clearly the queries the former theory leaves unanswered.

So, and generalizing this case, I believe that the "theory of the chain reaction" that I discuss in chapters six and seven, provides an answer to many of the questions that the already existing theories cannot answer; I also feel that the "theory of the chain reaction" will decisively contribute to the establishment of the sought for theoretical model that will explain in a more convincing and understandable manner what happens in the Universe. The "theory of the chain reaction" starts with the creation of the elementary particles-charges, which are charges without mass or dimensions, and then, through clear and simple arguments, comes up to the completion of the creation of our Universe, the other Universes and Antiuniverses and then that of the whole Cosmos, providing a complete picture of the creation and explaining all the developments without leaving any serious obscurities, unsolved questions or big gaps.

According to the "theory of the chain reaction", the Universe, which moves and evolves in a way within the "cosmogonic gas", as described in chapters six and seven, becomes increasingly larger as new quantities of matter and antimatter are produced in the cosmogonic gas; and this in quantities increasing with the rate of a geometric progression. The quantities of matter and antimatter produced supply the Universe and the other Universes and Antiuniverses that evolve together with our Universe. Thus, the "theory of the chain reaction" becomes the first theory that explains in a clear and grounded manner what happens to the antimatter that is produced with matter.

Along with the development of the Universes and Antiuniverses, a new concept is created, that of the "Cosmos". From now on we shall use this concept in order to refer to the whole of the Universes, Antiuniverses, areas of cosmogonic gas and all the void and completely void spaces that exist and that were and will been for ever infinite.

Until now, when we refer to the Universe, we mean large or very large dimensions. From now on, when talking about the dimensions of the Cosmos, our thought has to become familiar with huge dimensions compared to our way of thinking. These dimensions are exponentially bigger than those of our Universe and approximate the notion of infinity or of a gradation of infinity, if, of course, we can accept or assume that the notion of the infinite can have approximations or gradations.

I had to mention the above about the evolution of the human knowledge about the world of infinity until now. However, let us leave the research about the world of the infinite and see what happens with the evolution within the microcosm that means, with the evolution inside the molecule and the atom and inside the nucleus of the atom.

THE EVOLUTION OF THE KNOWLEDGE ABOUT THE MICROCOSM

The evolution concerning the "megacosm" and the "cosmos of the infinity" may have been very important, but the evolution concerning the microcosm has been even more important. Since the beginning of the past century, a big race for the study and the description of the microcosm has started.

At first, the structure of the atom and its nucleus was studied and it was discovered that the elements that compose the atoms and their nuclei are the well known now to all of us particles electrons, protons and neutrons, which, initially, until some years ago, were considered to be the elementary, indivisible particles from which matter and material bodies were formed.

Then, the behavior of the particles inside the atom was studied and the fission of the nucleus of the atom was achieved. At the same time, two new interactions were discovered from nuclear fissions and transmutations of atoms; these are the cause of the strong and the weak nuclear forces. A multitude of particles were also discovered, each of which plays its own role in the formation and the evolution of the Universe. At the same time, radioactivity was discovered and the study of the behavior of subnuclear particles started.

Very recently it was found that protons and neutrons, in their turn, are also divisible particles and are structured of other smaller particles. The particles that constitute the protons and the neutrons were name "quarks". At this moment, science is trying to find out what happens with the quarks and whether they are in turn divisible or indivisible particles. Concerning this issue, we have solid indications that the quark is a divisible particle, as

I describe analytically in the respective chapter. There is also evidence that leads to assume that electrons are divisible particles too.

New sciences, such as atomic and nuclear physics, were created for the study of the above and new theories were developed, as the "theory of quanta", QT and the "theories of relativity", SR and GR. At the same time, the old instruments and the simple experimental devices were replaced by high precision electronic appliances, powerful reactors and accelerators. All those facilitated the study of many issues and resulted in thorough progress in the research of the microcosm.

Certainly, despite the significant discoveries in the microcosm, the questions that remain unanswered are still many and very basic questions. For example, I mention that we know very few things about the behavior of the subatomic particles inside the nucleus and even less about the way in which energy is transformed into matter and the details of this transformation; or about the origin, the nature and the behavior of the fundamental interactions –forces–. Especially, our knowledge about the nature and the origin of the gravitational force, which is the force that connects the microcosm –micro particles, atoms and molecules– with the megacosm, –mass, matter and Stars– and the world of the infinite, –Solar systems, Galaxies, Universe, Universes and Antiuniverses– and its relation to the electromagnetic force and the other interactions, remains incomplete.

Einstein, feeling the deficiencies created by the above obscurities, after formulating the "special theory of relativity", tried to provide several explanations mostly for the case of the discovery of the cause of the force of gravitation and its unification with the electromagnetic force, as he believed that by giving correct answers for this unification he would resolve many of the above obscurities. So, in 1916, Einstein envisaged the new theory, the "general theory of relativity", which he believed would answer the above questions; however, the formulation of the theory did not yield any substantial results. This happened because the "general theory of relativity" was limited in only theoretical results, with complicated mathematical expressions which once more did not provide any substantial facts about the real cause of gravitation. Therefore, it was considered that the general theory of relativity rather perplexed things further and for this reason it was gradually abandoned.

Although, from 1930 until the end of his life, on April 18, 1955, Einstein, made a lot of efforts to discover the cause of gravitation and its unification with the electromagnetic force, unfortunately, this goal proved very difficult even for a personality like Einstein. During the period from 1960 to 1975 the interest of many scientists in the general theory of relativity came back to the forefront. Lots of papers about the theory were written and published, but all stop being understandable from the moment they start to describe something important about the theory. At the same time, all those efforts did not achieve to describe the causes of gravitation, but dealt once more with the mathematics and the theoretical analysis of the gravitational field rather than the cause of gravitation.

Many new researchers, perhaps for their personal prominence, are presented as followers of Einstein's work, trying to combine the general relativity with the quantum theory, in the field of gravity, hopping that this will lead to some positive results. Unfortunately, however, these researchers have not realized that, even if they do achieve a combination, it is rather impossible, that this combination can lead to the cause of gravitation[1].

Completing the description of the advances concerning the microcosm and changing subject, I would like to note that the phenomena in the microcosm and the world of the infinite, contrary to those of the megacosm, are complex and at times unforeseeable, complicated phenomena permitting multiple and different explanations. In some cases, these explanations reach even fictional, extreme, non scientific views. In this case, I feel that a lot of caution is needed as there is a risk for physics to be turned into a set of fictional and unrealistic descriptions. My personal view is that those cases should be evaluated with special caution and be

---

[1] See my work. A New Proposal that Completes and Verifies, Newton's Gravity and the Law of the Universal Attraction: viXra:1905.0193.

incorporated in science only when the scientific society is convinced that they are correct and express indisputable scientific ideas indeed.

So this is the evolution in human knowledge about the "microcosm", the "megacosm" and the "cosmos of the infinity" until now. This is why I believe that this book and then the whole the "Trilogy of the Creation" will offer in their turn a small, very small contribution so that physics continues unobstructed in its path to the correct and creative development with indisputable facts and correct arguments.

Chapter 3

THE UNIVERSE AS IT IS KNOWN TODAY

If for a moment we ignore the details about black holes, cosmic tunnels, quasars, giant stars, dwarf stars, etc., a general picture of the Cosmos according to the current knowledge leads us to a Universe consisting of a vast number of stars and void spaces, which form the solar systems. The solar systems, in turn, form a very large number of galaxies, which, all together form the Universe. But is this picture we know today the definite one of the Cosmos?

According to the "theory of the chain reaction", the Antiuniverse and probably other Universes and Antiuniverses are formed along with the Universe. So might "Cosmos" be much bigger than we think. And might its picture take a form closer to that suggest in Figure 20 page 117.

THE SKY AND THE STARS

If we look upwards in daylight, we see the sky which has a beautiful blue color. Above it there is an extraordinary bright diamond, the Sun, which, as we all know today, is one of the many stars in our Universe. But as the Sun is very close to us compared to the other stars, we use to call it "our Sun".

During the night, the blue color of the sky becomes black; however, the sky, despite its black color in night remains beautiful as it is adorned by lots of small lights, which we all know as stars. It is also adorned by another, bigger light, the Moon. Of course, science is aware of the fact that stars exist in the sky in day too, but they are not visible because they are covered by the Sunlight which is brighter than the light of the stars. The void space between the stars is the space that from what we know is boundless.

Have you ever wondered what would be the color of space? If someone asks you about the color of space, do not hasten to say it has the color of the sky –that it is blue– or that it is blue during the day and black during the night, because you will be certainly wrong. Space is always black. However, during the day, when the sunlight meets the molecules of the atmosphere of the Earth, it falls on them and is dispersed towards all directions, producing a bluish bright glow in the sky. So the sky has the blue color only in the regions of the atmosphere that are near the Earth and "seen" by the sky.

THE OBSERVERS OF THE SKY

The first research about the sky was performed with our eyes, which were the first instruments people used for their observations. However, human curiosity for further knowledge about the stars and what else might be hidden in the sky led the researchers to build the various instruments with which they tried to collect more details from their observations.

Today we use thousands of instruments in the study of the sky and the stars; however, the description of these instruments does not fall into the scope of our work. In this section, we are just going to describe very briefly two of those instruments, the "telescope" and the "spectroscope", which are two of the most important instruments used for the observation of the sky.

We proceed to this very brief description of the instruments in order to give the reader the chance to understand the way in which physics performs its observations and calculations in order to reach several conclusions, as for example, that the sky does not contain six thousand stars only, which are the ones we can see with naked eye, but several billions of stars and that besides our galaxy the Universe consists of millions of other galaxies. With these instruments we can also study the materials a star is made of, whether it moves, its speed, etc.

The first of these instruments, the telescope, is an instrument that, through a system of lenses, brings the various objects closer magnifying them at the same time. So, if we observe the sky through a telescope, we shall see that it does not contain only the stars seen by naked eye, but millions of stars. We shall also see a lot of details of the various stars and note the thousands of galaxies in the Universe.

However, human efforts to yield increasingly more information from the sky led to improved telescopes by the replacement of the lenses with mirrors. So the "reflector telescope" was built, in contradistinction to the initial one that was called the "refractor telescope". The reflector telescope provided more information. Then, the radio telescope was invented, which remains the most perfect instrument for the observation of the sky.

Another instrument is the spectroscope. The spectroscope is a very simple instrument that consists of two lenses and a prism and analyzes the light of a light source after bringing it closer; it provides its spectrum, which we could briefly say is the "set of colors consisting the light of the source". As strange as this may seem, we can yield thousands of pieces of information about a light source from this spectrum. Therefore, by observing the spectrum of the light of a star, we can learn whether this star is drawn away or approaching the Earth; the speed at which it moves if it approaches or is drawn away; its temperature; its materials; its age; and a lot more information, as we shall see next.

The development of the spectroscope had as result the "spectrograph", which is in fact a spectroscope in which a photographic chamber is incorporated. So we can photograph the spectrum of a light source and then study it better.

WHAT IS A STAR?

So, with the help of the telescope, the power of which corresponds to about "one million human eyes", the stars that we see during the night with the naked eye as mere spots in the sky become big balls and come much closer. We can distinguish a lot of colors on them: light blue, blue, white, red, yellow, green, etc.

On several of the closest stars we can even distinguish several shapes of mountains and plains or clouds of gases, etc. We can also observe, if we watch more attentively, that the stars remain still in the sky. But some of them seem to move among the other stars that remain still. So we can divide the stars in two large classes, the planets and the fixed stars. Usually, the fixed stars, such as the Sun, are "self-luminous", they have their own light which is produced by the high temperature on their surface, whereas the planets, as for example the Earth, are non-luminous and therefore they are illuminated by other stars.

Of course, we now know that all stars and planets and fixed stars "move". However, previously, as the fixed stars due to their distance seemed still in the sky, people thought that they do not move and therefore they called them fixed stars.

After many years of observations with the telescope, astronomers found initially that there are not only the six thousand stars that can be seen with the naked eye in the sky, but more than twenty thousand stars. Then, with the invention of the radio telescope and the photographing of the stars, we realized that there are not only twenty thousand stars in the sky but that the existing stars are more than three billion! Now, according to the "theory of the chain reaction", this number of stars in the sky might increase to inapprehensible for the human mind.

In general, stars are material bodies and are all different. Others are very big, probably even millions of times bigger than the Earth or the Sun, and others are much smaller. Others are in solid state, others in liquid and others in gaseous state. Their temperatures are different, from some degrees below zero in certain stars to many millions of degrees above zero in others.

After many years of observations and laborious efforts, astronomers classified stars in groups that include also the well known to all constellations of the zodiac. Thus they drew the charts of the sky separately for the northern and the southern hemispheres and made lists with the millions of stars to which new discoveries are continuously added.

Gradually, they calculated the distances of many stars from the Earth. In order to be able to express these distances, which were enormous, they established new units for the measurement of distances, such as the distance of a light year, that is the distance light travels in a period of one year, i.e., nine thousand four hundred and sixty billion kilometers per second.

So it was measured that the Moon is 384,000 kilometers away from the Earth, the next closest star of our solar system, Mars, is 56,000,000 kilometers away, the Sun 150,000,000 kilometers and the farthest star of our solar system, Pluto is 5,000,000,000 kilometers away from the Earth. Further than the stars of our solar system, which are considered to be relatively close, one of the next closest stars is Sirius, which is eight light years away from the Earth, whereas Vega is 22 light years away and the stars of Orion are 1,500 light years!

We all know that the bigger a light body the more light it emits around it. Does this also apply to the brightness of the stars? Certainly not, because, when we observe a star from the Earth, its brightness does not only depend on its size but on another factor as well, the distance. So, a star may be seen brighter than another star, not because it is bigger, but because it is closer to us. Based on a correlation of the brightness and the distance of a star we can then calculate its size.

Now, in order to form an idea about the size of the stars and their variety, let us proceed to the following comparison between stars –a comparison Astronomers use–. We shall classify the stars according to their size into two large classes, the "giant stars" and the "dwarf stars".

However, despite this classification, the variety in the size of stars is such that scientists established once again a new unit in order to be able to classify them. This unit is the size of our Sun. So, we can say that a star is up to one thousand times bigger than the Sun; up to this size, a star is classified as dwarf! Or we can say that a star is one thousand times bigger than the Sun; from there on, we classify the stars as giants.

Certainly, one might think if it is possible to find even one star a thousand times bigger than the Sun. Such a thought, however, is mistaken, as there are not only innumerable stars more than a thousand times bigger than the Sun, but there are stars even one million times bigger than the Sun!!! Astronomers named those stars after the specific name "supergiant stars".

So stars up to one thousand times bigger than the Sun are called dwarf stars. From then on, stars up to one million times are called giant stars and then they are called supergiant stars. According to this classification, we may say that our Sun is a dwarf star!

Two other classes of stars, the names of which we find very often in cosmological textbooks are "nova stars" and "supernova stars"; these names are related to the Latin word nova, which means new, and in this case it means new star. Nova stars owe their name to the fact that those who studied them noted that sometimes they were not visible at all and sometimes they were very bright, visible even with the naked eye. So the astronomers that observed them thought they were new stars and therefore they named them nova. Several of them were so bright and impressive indeed, that they were named supernova.

In fact, these stars are not new, but they were dwarf stars, as described above, which due to explosions inside them grew bigger and from dwarf stars they became giant stars. The cortex of those stars is overheated due to their high internal temperatures and starts to radiate; so, although these stars are enormously far from the Earth, they become visible stars. These stars are given the specific name red giants. The explosion of the red giants forms then the other stars which gradually give the current form of the Universe, as described in the next section.

With the use of the spectrograph and the spectroscope we learned about the materials the stars consist of, the way in which they move, the speed of their movement and whether they approach the Earth or are drawn away from it. By analyzing the spectrum of the stars, we analyzed their components and we drew information that showed that these constituents are –proportionally, of course– the same in the whole Universe. Among these components, the elements Hydrogen (75%) and Helium (23%) prevail, whereas all the other constituents in our Universe do not cover more than a mere percentage of just 2%. So this is the proportion of the elements in the Universe; scientists named it "cosmic proportion".

Certainly, after reading the above paragraph, you wonder how it is possible to have in the Earth, as we all know such different proportions of the elements. This happens because the above proportions are calculated in cosmic scale. However, in smaller scales, there are Hydrogen stars, which are named like that due to the fact that in those stars the percentage of Hydrogen prevails; Helium stars, where the percentage of Helium prevails; Calcium stars, where calcium prevails, etc.

Anyway, I feel I have fairly tired you with the descriptions of the stars and that I have slipped away from the basic subject of the study. For this reason I put a full stop in the description of the stars, noting that, in general, there is a multitude of more detailed textbooks available for those interested in learning more about the stars. In this section we tried to describe the stars in general and provide the reader with just an idea about what happens with the stars in the Universe, thinking that in this way he/she will better understand what we analyze later.

"Quasars", "Red Giants", "White Dwarfs", "Pulsars" and "Black Holes"

Besides the stars we described in the previous section, there are also many other celestial bodies, which due to their peculiarity are described as separate entities and as separate phenomena of the Universe. In some cases, Astronomy grouped those celestial bodies giving them specific names. We shall describe some of those bodies in order to provide a general picture about what else exists or happens in the Universe and be able to distinguish them when we find them in various Cosmology textbooks.

So, one group of those celestial bodies is that of the "quasars", which were discovered in the early 1960s. Quasars are very bright celestial bodies with the size of a galaxy, visible by telescope. They are the primary protogalaxies and emit huge quantities of energy around them. They are very far away, almost at the edge of the Universe. A similar, parallel name for the quasars is radio galaxies.

It is believed that the furthest quasars of all represent galaxies in their initial stages of creation. In order for you to have an idea about the distance between the quasars and the Earth, we state that quasars were located at a distance of 10 billion light years!

I read in a Cosmological textbook the following remark, "several authors wonder whether the quasars are concentrated matter and antimatter" and then they question "whether the quasars are the boundaries of the Universe or is the Universe even bigger?" It was the first time I read something about antimatter connected even faintly to a theoretical aspect. At that point I thought or rather I felt that the "theory of the chain reaction" provides answers quite to the point in what concerns the above questions, proving that, indeed, large concentrations of matter and antimatter are formed at the boundaries of the Universe and that they are not yet completely separated.

Another class of celestial bodies with the size of very big stars is that of the "red giants". The red giants, as mentioned in the previous section too, are supernova stars which, due to nuclear reactions that take place in their center and release large quantities of energy, are excessively enlarged. However, as millions of years pass, the inner part of those stars gets filled with nuclei of carbon and many new elements.

These stars explode in the end and their matter is extinguished at very high speeds to huge distances. This development becomes the starting point of new cosmic developments through which the solar systems, the galaxies and the Universe are formed as we actually know them. It is supposed that our solar system originated also from such an explosion of a red giant.

However, sometimes it happens that red giants use up their nuclear fuels before they explode. In this case, they start to shrink and, according to their mass before using up their fuels, they turn gradually into white dwarfs, pulsars or black holes, as we describe later.

If the red giants that do not explode have mass bigger than one and a half times the mass of the Sun, they shrink and form very small stars with a diameter of just several thousands of kilometers. Astronomers call these stars "white dwarfs".

Now, if the red giants have coincidentally about the same mass with the Sun, then stars even smaller than "white dwarfs" are formed, with a diameter of about 15 kilometers. These are the "neutron" stars or "pulsars". Physics accepts that neutron stars are mixtures of neutrons, protons and electrons and that the repulsive forces of neutrons are balanced by the attractive forces according to Pauli's exclusion principle. At this point, the "theory of the chain reaction" foresees pulsars and gives a completely different explanation about those stars, i.e., that pulsars are large concentrations of neutrons and protons, like nuclei, around which electron clouds spin. That is, they are something like huge individual atoms. The coherence of protons and neutrons in pulsars, according to the "theory of the chain reaction" is based on the properties of the "strong nuclear force", which we describe in the third book, of the "trilogy of the creation", in the "formation of matter and antimatter".

Now, in case that the mass of a red giant is bigger than the marginal mass for the creation of a white dwarf or a pulsar but this red giant does not explode, we may suppose that a star with very powerful gravitational attraction is formed and that this gravitational attraction is so powerful that it even modifies the paths of the light rays it emits. So the light the star tries to emit cannot get away from its surface, because it is attracted by its gravity and turns back. The regions where such concentrations are found were named "black holes".

Although black holes has not been experimentally found yet, they are traced indirectly due to their intense gravitational effect in the surrounding space. It is believed that there are lots of black holes dispersed in the Universe, mostly at the centers of the galaxies.

THE SUN THE EARTH AND OUR SOLAR SYSTEM

Previously, in the first section of this chapter, we called the Sun "our Sun". We did this because the Sun, compared to the other stars, is very close to us, and so, as we shall see later on, together with the other stars, the "planets", which are also close to us, we belong to the same stellar neighborhood and the same stellar system, our "solar system". So, certain stars, the "planets", namely Mercury, Venus, Earth, Mars, Jupiter, Saturn, Uranus, Neptune and Pluto, move around the Sun at fairly short distances compared to those of other stars, our galaxy and our Universe. Often, other stars, called satellites, move around the planets. So, the Moon, Earth's satellite, moves around it, three satellites move around Saturn, etc.

For informational reasons only, we shall provide a very short description of the stars of our solar system starting from the Sun, which is the center of our solar system and continuing our description up to Pluto, the furthest star of our neighborhood; although we have been aware of Pluto's presence since 1905, it had not been located until 1930.

Thus, we could say that the well known to all Sun, the star at the center of our solar system, is a star singled out from the other stars of our neighborhood due to its very large mass and its high temperature. Its large mass is the reason that keeps it at the center of our system. The very high temperature inside the Sun, that reaches approximately 20,000,000°C and is due to the nuclear fusion of hydrogen to helium, is what forms its outer layer, the photosphere, with a temperature of 6,000°C, the main reason of its light radiation.

It is a third generation star, as Astronomers call it, and was formed by other stars explosion debris. Part of the material of the explosions formed the Sun and several other small quantities of heavier elements formed the planets that move around it. As it is not of female gender, I believe it will not matter if we mention its age, which is about six billion years; and in case someone worries about its future or its health, we would like to inform him/her that under normal conditions the Sun will live at least five billion years more.

Mercury is the closest planet of our solar system to the Sun. It has a diameter of 4,720 kilometers and it is the smallest planet. It is 60,000,000 kilometers away from the Sun and makes a rotation around it in 59 days. The temperature on its surface ranges from −170 to +350°C. The conditions in its atmosphere are such that render the existence of life on the planet impossible.

The spacecraft Mariner 10, which approached the planet in 1974, discovered a world similar to the Moon. Mercury is the fastest moving planet in our Solar system.

Venus is the second closest planet to the Sun. Many times it appears in the sky after sunset or early in the morning before dawn; then, depending on the time it appears it is called either Evening Star or Morning Star. It completes a rotation around the Sun in 226 days, at an average distance of 108,000,000 kilometers from it. Its diameter is 12,370 kilometers.

In 1962, the spacecraft Mariner 2 passed by Venus at a distance of 20,000 kilometers and disclosed that it is a very inhospitable planet. It has very high atmospheric pressure of about 100 atmospheres, a fact meaning that its atmospheric pressure is 100 times higher than that of the Earth; its surface, where the temperature reaches about 480°C, is like a sea of half melted rocks.

Venus' surface is surrounded by a dense atmosphere, 100 kilometers thick, consisting of carbon dioxide, nitrogen and sulfuric acid. It has huge mountains and big volcanoes. Its atmosphere is swept by heavy storms, tempests and lightning. In 1967, the spacecrafts Venus 4 and Mariner 5 visited Venus again and confirmed the results of the visit of Mariner 2.

Our Earth is the third star closest to the Sun. It is the star that hosts the characteristic of "Life", which means that it is inhabited by creatures realizing what happens in the Universe. The Earth is probably the only star in the Universe where life exists. This presumption results from the fact that no other star has been discovered yet with the characteristics or similar conditions that might host living organisms.

The Earth was formed along with the other planets of our solar system 4.6 billion years ago. At first, the Earth was an incandescent mass sustaining a shower of meteorites. Gradually, however, it started to cool down and freeze and its surface solidified and plains and mountains formed. At the same time, its first atmosphere was formed, consisting of gases that were released from the incandescent masses. The strong storms and the heavy rains that followed resulted in the filling of the vast cavities that had been formed with water, thus forming the lakes and the seas. The basic geological modifications of the Earth were completed about 500,000,000 years ago.

However, the most important event, which might also be the second in significance among the events of the creation of the "Cosmos" after the creation of the elementary particles and the interactions is the formation of compounds that constituted the starting point for the creation of "Life", which appeared in the beginning in the form of unicellular organisms that then evolved into the current levels we all know.

The Earth has only one satellite, the Moon that moves around it and is at an average distance of only 386,000 kilometers. The Moon, with a diameter of 3,480 kilometers moves around the Sun with the Earth and around its axis in 27.3 Earth days. The rotation of the Moon around the Earth is the same with its rotation around its axis, a fact that makes it always have its same part turned towards the Earth. The gravitation force on the Moon is very weak and as a result it has no water or atmosphere.

The moon was the first goal of humans in the exploration of space. The conquest of the Moon took place on July 20, 1969 with the spacecraft Apollo 11. The conquest of the Moon was a triumph for humans and their technology, as it verified the laws of nature science had predicted. The conquest of the Moon verified experimentally too that humans can live and work comfortably at weak gravity, as that of the Moon.

The next planet in our solar system is Mars, known also as the red planet. Its average distance from the Sun is about 220,000,000 kilometers and its closest distance from the Earth is 56,000,000 kilometers, something that happens every 15 years. Mars' diameter is 6,750 kilometers; about half of the diameter of the Earth, and its rotation around the Sun lasts about 687 days. It rotates around its axis in about 24 hours, with about the same rotation period as that of the Earth. It has two satellites in total.

Many things have been written about the possibility of the presence of life on Mars, but the information collected in 1965 by the spacecraft Mariner 4 was disappointing. The photographs Mariner 4 sent showed that craters are dispersed all over the planet; it has a thin atmosphere, without oxygen, which reaches up to 40,000 meters in height; it is cold and temperatures from -7 to +10°C prevail. Dryness predominates of the surface of the planet and no traces or conditions favoring the presence of life were found.

The spacecrafts Viking 1 and 2 that landed on Mars in 1976 absolutely confirmed the above results. However, many people believe that the red deserts of Mars are vast areas of iron oxides that might correspond to a Cosmos which has been destroyed and deserted by oxidation and that probably a civilization existed in the past and then vanished.

After Mars a sudden change is noted in the rage of distances in our Solar system. There is an immense, vast void of about 45,000,000 kilometers and then the planet Jupiter follows, which is the biggest one in our solar system. Its diameter is 11 times bigger than that of the Earth and its day lasts about 10 hours. Its rotation around the Sun lasts about 12 years. The temperature on its surface is –130°C.

Jupiter's gravity is two and half times that of the Earth's and its atmosphere is manifested as an ocean of clouds creating very high atmospheric pressure. As the largest part of its matter consists of hydrogen and helium, it is considered the star that failed to be turned into a Sun. In case Jupiter turned into a Sun, the consequences for our region could not be known.

Jupiter has 16 satellites. Of special interest is Io, which was discovered in 1610 by Galileo. Io has a diameter of 3,630 kilometers and is a satellite full of craters and active volcanoes. The satellite Ganymede is also of interest; it is considered the biggest satellite in our Solar system, bigger than the planet Mercury.

After Jupiter, the second biggest planet of the solar system, Saturn, can be found and in extreme cold conditions. Saturn's rotation around the Sun lasts 29.5 terrestrial years, whereas its full rotation lasts 10 hours and forty minutes.

The system of multiple rings surrounding the planet is one of its characteristic features. This system consists of seven basic rings extending many kilometers beyond the planet. These rings consist of millions of small satellites formed of sand, dust and ice. Saturn's temperature is about −200°C. It has 17 satellites in total.

Next comes, the Uranus, which was discovered accidentally in 1781. It is 2,873,000,000 kilometers away from the sun and rotates around it in 84 years and 7 days. Its diameter is 50,500 kilometers and its volume 63 times bigger than that of the Earth. Uranus has several thin rings too and 15 satellites.

Its surface is steep, with precipices up to five kilometers. The Voyager 2 approached Uranus in 1986 and gathered valuable information about the planet. The radio waves Voyager 2 emitted traveled two hours and forty minutes before they reached the Earth.

Neptune is the next, penultimate planet in our Solar system. It was discovered in 1846. It has a diameter of 57,000,000 kilometers and it takes about 165 years for a full rotation around the Sun. Voyager 2 approached Neptune in 1989 and showed, among other things, five rings and eight satellites. One of its satellites, Triton, rotates inversely than its other satellites.

The planet's color is light blue, its atmosphere consists of helium and methane and its surface is overrun by cyclones and storms. A huge dark spot approximately the size of the Earth appears at its center. Its temperature is about −240°C.

With the description of Neptune we reach the last planet of our solar system, Pluto. This was discovered in 1930. Its diameter is about 6,300 kilometers and is volume about one tenth that of the Earth. In 1978 its unique satellite, Charon, was discovered,

Let us leave now the description of the classes of big stars and study several classes of smaller celestial objects that contribute in turn to the configuration of our solar system and the Universe. A class of such objects is that of "comets".

Comets are celestial bodies moving in elliptical orbits around the Sun and travelling huge distances. This is the reason for their long term absence that can sometimes last even hundreds of years. During their journey in the solar system, their orbits cross the orbit of the other planets and thus, when they approach the orbit of the Earth, we see them for a while and they disappear again. Comets are also the origin of the popular saying, "he passed like a comet", about people that pay rare and short visits to relatives or friends. Some of the comets appear at regular intervals, but there are also many comets that appeared just once and have never appeared again since then.

Comets move at an average speed of about 60,000 kilometers per second and have a different speed when they are close to the Sun and a different one when they are very far from it. They consist of concentrations of very hot luminous gases of very low density. The dimensions of the comets are approximately the same as those of the Earth.

The collision of a comet to a planet takes place in an intense explosion wave. The collision and the explosion produce high heat and a strong flush. It is said that part of a comet fell on Siberia in the summer of 1908 causing a terrible explosion and great devastation in a region of one billion square meters.

Another class of celestial bodies that contribute to the configuration of our solar system is that of the meteors. Meteors are very small bodies, with sizes that start from that of a rock and reach that of a small grain of sand or a pebble. There are billions of such bodies rotating around the Sun, the planets, even around the satellites of the planets. They are the small luminous spots we have all observed entering the atmosphere of the Earth and then disappearing over the horizon. Due to the fact that when meteors enter the atmosphere they lighten like stars, scientists often call them also shooting stars.

The above classes of comets and meteors are completed by the asteroids, the parasites of the sky as these celestial bodies are usually called. Asteroids are very small planets, with a diameter of only several kilometers, moving in the huge void we discussed previously, that exists between the planet Mars and the planet Jupiter. It is estimated that about 100,000 asteroids are in that void.

OUR GALAXY AND THE OTHER GALAXIES

We now know that there are millions of other suns and millions of other solar systems in the Universe besides "our Sun" and "our solar system", which we described in brief in the previous section.

At nights, we can observe in the sky, the stars and, on cloudless summer days, an expansive light; something like a cloud that astronomers found consists of millions of stars, which are so far from us, that are not seen clearly. The stars that form the image of this cloud, Cosmology named galaxy. We know today that our solar system is inside this galaxy. And, as in this way we are a small part of this galaxy we use to call it, like our solar system, "our galaxy".

The stars in our galaxy have a spiral distribution in the form of a disc with a diameter of one hundred thousand light years. The younger stars are in the spiral branches of the disc where the bigger quantities of hydrogen are concentrated. Our solar system is approximately two thirds between the center and the outer end of the disc of the galaxy.

Besides our galaxy, there are also thousands of other galaxies in the Universe. One of those galaxies is visible from the northern hemisphere with naked eye. This galaxy, which is at a distance of about two million light years from us, was named Andromeda nebula. The closest neighbor to our galaxy is the Magellan nebula, which is visible in the sky of the southern hemisphere and is at a distance of two hundred light years. In order to distinguish them, Cosmology gave various names to many galaxies, as for example, the two galaxies we mentioned above, the "a" of the Swan, the nebulas in Cancer, etc. Depending of their shape, scientists classify galaxies in various types. So we have spiral, elliptical, rod-like galaxies, etc., however, as many scientists believe, they all end up in a round mass which is the final stage of their development.

THE UNIVERSE

The billions of stars in the sky, the millions of solar systems, the thousands of galaxies and the vast voids between all these form our actual Universe. In order to provide an idea about the greatness of the Universe, we give some numbers related to the Universe:

The age of the Universe, according to the theory of the "big bang" – which is the most prevailing theory about creation–, "from the moment of the big explosion until now" is estimated to be at least eighteen billion years.

The diameter of the visible Universe is estimated as $3,6 \times 10^{10}$ light years distance. "The range of a light year is the distance light travels in a period of one year, i.e., about ten trillion kilometers". Let us give the diameter of the Universe, which is 35,000,000,000,000,000,000,000 meters in order to realize the scale of numbers we are talking about. The mass of the visible Universe is estimated as 10^{55} Kg.

The above numbers are huge and rather inconceivable compared to the compatible dimensions that are used in our measurements. However, these dimensions do not end at the inconceivable numbers of the Universe. In the "theory of the chain reaction" we see much bigger numbers, concerning the Cosmos, which are exponentially bigger than those of the Universe, reaching infinity.

The Cosmos, the Universe, the Universes, the Antiuniverses and the Theory of the Chain Reaction

All those we discussed in this chapter are a very brief summary about what the Universe is and what happens inside it. In this way the range of human knowledge about the Universe is completed; in general, we might say that this knowledge starts from the "nucleus of the atom" and reaches the "boundaries of our Universe".

However, as these boundaries are closed boundaries in knowledge, two new basic questions rise; the first is what happens inside the "nucleus of the atom" and the second is what happens with the "boundaries and the development beyond the boundaries of our Universe". Thus, the "theory of the chain reaction" that we describe in chapters six and seven of this book, enlarges the boundaries of our knowledge, on one hand about the nucleus, starting from the elementary particles charges, "pointons" and "antipointons", giving the atom its new form as shown in Figure 19, page 107, and going further to infinity, creating the concept of the "Cosmos", which includes the "Universe or the Universe" and the "Antiuniverse or Antiuniverse", the "cosmogonic ga" and the regions of "absolutely void spaces" that reach infinity, as shown in Figure 20, page 117. So, the "theory of the chain reaction" guides us and brings us probably closer to the definite image of the "Cosmos", an image that starts from zero, represented by the "particles pointons and antipointons, which are particles without mass or dimensions" and goes further to infinity, represented by the inexistent "boundaries of the absolutely void space".

Chapter 4

MATTER, ELEMENTARY PARTICLES AND FUNDAMENTAL FORCES

> Matter consists of just a few fundamental elements, six quarks and six leptons. And for its structure it uses even fewer of those elements, "just two of the quarks and the electron". It is difficult to conclude whether this picture is the definite one. As long as the interactions between particles are discrete and only four, the picture seems –at least for the moment– simple and symmetrical again. What remains to be cleared, in order for us to understand the Universe, is in what manner these particles interact both between themselves and with the forces we have already mentioned.
>
> *George Grammatikakis*
> From the book *Veronicas' Hair*

ABOUT MATTER

The word "matter" defines the substance, from which the various material bodies are structured. Matter can be conceived by a series of its properties, such as weight, mass, volume, shape, etc.

Initially, the notion of matter was quite simple for human beings, as their conceptions about matter were confined to the description of the way man made from it the various physical objects he used for the needs of his life.

However, gradually, as his thinking evolved, man posed several questions such as, which are the elements that created matter; how and why matter deforms, or why it appears in different states, etc. Those questions were the cause that led to the study of the philosophical concept of matter and research, of its nature and structure in general.

Among the first people that studied the philosophical concept of matter where the ancient Greek philosophers; among them the names of the Pythagoreans, Thales of Mellitus, Anaximander, Heraclitus, Zeno of Elea, Diogenes and many others are distinguished, having each interpreted in his own way the behavior and the entity of the material objects.

Aristotle's points of view were remarkable, as he believed that matter within the Universe consists of four elements, earth, air, water and fire. Two forces were exerted on those elements. Those forces were gravity, which was expressed by the tendency of earth and water to move downwards, and lightness, which was expressed by the tendency of air and fire to move upwards. It is worth noting that this distinction, among the components of the creation, the evolution and the function of the Universe, is used even today.

According to Aristotle matter is continuous. This means that we might divide a piece of matter in smaller pieces limitlessly. By this notion, there is a continuous division of matter, which means that we shall never find a fragment of matter that we shall not be able to divide in smaller pieces.

On the other hand, Democritus maintained the opposite, i.e., that matter is granular by its nature and that every material body consists of a very large amount of different atoms. And in this case, with the word "atom", Democritus described precisely the last subdivision of matter; a subdivision after which, matter could no more be further fragmented.

Research and discussion on Aristotle's and Democritus' various points of view, lasted many centuries and neither part was able to present either a theoretical or an experimental proof confirming the correctness of its arguments. This lasted until 1803 when Dalton, the famous British physicist, in his effort to explain the phenomenon of multiple proportions of the elements in the various chemical reactions, formulated the thought that this phenomenon is due to the fact that matter consists of small particles which he named "atom", using the same name that Democritus had used twenty centuries earlier.

The above conflict between Aristotle's and Democritus' points of view, as they have been supplemented by Dalton, continued for one more century, until 1905, when Einstein[2] made a very important observation about the existence of the atoms.

Einstein's observation was that, within a liquid or a gas, a random continuous movement of various dust particles is performed. This movement, –which physics has named "Brownian movement", after the scientist who observed it– as a physical phenomenon, could be explained only by the movement of the "atom" into the liquid or the gas and their collision with the micro particles of dust. Thus, based on Dalton's observations about the phenomenon of multiple proportions of the elements in the various chemical reactions and then on those of Einstein about the "Brownian movement", the experimental proof that matter consists of various small particles, the "atoms", which at first were considered also as elementary particles, i.e., as the last subdivisions of matter, was at last also provided.

THE ELEMENTARY PARTICLES

However, it seems that nature has a kind of peculiar and very reserved and slow rate by which it discloses us its secrets, forcing us to proceed to perennial research and quest. Thus, after the discovery of the atom, new questions started to arise about what an atom is, what it consists of and how it has been formed.

---

[2] In 1921 Einstein was awarded the Nobel Prize for his services to Theoretical Physics, and especially for his discovery of the law of the Photoelectric Effect.

At that time, around 1910, after the experimental proof of the existence of the atoms, suspicions that the atoms in turn should not be the elementary particles of matter, but must consist of other smaller elements. Already existed, several particles negatively charged, with a very lower mass – about one to two thousand times–, the well known today "electron" had already been experimentally traced in the mass of the hydrogen atom. Thus, the discovery of the electrons led scientists to the indisputable conclusion that other, smaller, subdivisions of matter also exist and that, therefore, its division does not end in atoms.

In 1911, Rutherford, the British physicist, proved that the atoms can be subdivided in small particles and, specifically, that they consist of a nucleus, which is positively charged, around which the electrons spin; the particles that had already been discovered. At that time it was considered that the nuclei of the atoms consisted of electrons and protons; protons being particles just as the electrons, but positively charged. They were thus named because it was thought that together with the electrons, consisted the elementary particles and the fundamental units of matter.

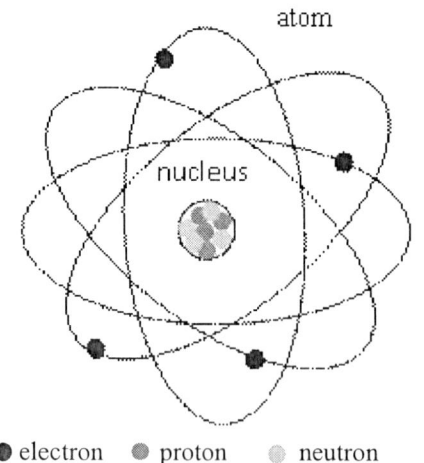

● electron ● proton ● neutron

Figure 8. The form of the atom after the discovery of the neutron.

This model for the structure of the atom, with the compact nucleus and the electrons spinning around it, did not last but just about 20 years, until

1932, when James Chadwick discovered that the nucleus of the atom, in its turn, does not consist of electrons and protons, but that it consists of two different particles indeed, with the following difference, however: one of those particles is the "proton" particle, which, as described above, had already been discovered, but the second one is not the electron but a neutral, new particle "without charge", with a mass approximately equal to the mass of the proton; this particle, due to the fact that it was neutral, was named "neutron". So the new image of the atom at that time was formed as shown in Figure 8, i.e., a nucleus, which is the compact part of the atom that consists of a total of protons and neutrons around which the electrons spin.

With the discovery of the electron, the proton and the neutron, which were considered the elementary particles of matter and after the completion of the above model for the structure of the atoms, scientists thought that research for the structure and the origin of matter had at last come to an end. Indeed, several scientists rushed to state that in a very short period of time research for the structure of the microcosm will have been completed. However, one more time, scientists were contradicted, as about thirty years ago, in experiments performed by Marrey Gellman, a physicist, at the California Institute of Technology, in which high speed protons collided with other protons, it was shown that protons and neutrons are not elementary particles. Specifically, it was shown that protons and neutrons are complex particles consisting of triplets of other particles, the "quarks". Then, several variants of quarks appeared, such as quark "up", "down", "strange", "top" and "bottom". Quark "colors" were also found, with variants as well: "red", "green", "blue", etc.

From those quark variants we shall take apart and examine the case of the quarks "up" and "down", which are the particles formed nucleons, i.e., the nucleus particles, the protons and the neutrons. The most important of the characteristic parameters of the above two particles, the quark "up" and the quark "down", are their mass and their charge. Both these particles, as physics accepts nowadays, have the same rest mass, $310MeV/c^2$, and as for the charge, the quark "up" has 2/3 of the charge of the proton and the quark "down" has 1/3 the charge of the electron.

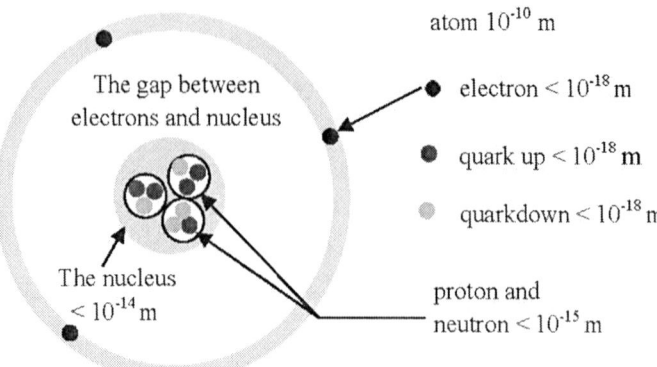

Figure 9. The form of the atom as of today.

The combinations by which protons and neutrons were formed from the quarks are: two quarks up and one quark down formed a proton. And two quarks down and one quark up formed a neutron. The way in which these combinations took place is shown in Figure 10, on the page 67.

Therefore, the image of the structure of the atom after the discovery of the quarks has the form that remains until now days, as shown in Figure 9, above.

A Very Basic Question: Are the Quarks –Up and Down– Elementary Particles or Not?

At this point and after the discovery of the quarks –"up" and "down"–, a great and very basic question rose again! Are the quarks elementary particles? That is, are they the smallest subdivisions of matter? Or does its division continue more, even after the quarks?

The experiments that have been performed until today have not found any other smaller subdivisions of matter after quarks. However, there is something else too; as the wavelength of light is bigger than the size of the atom and therefore bigger than the size of the elements consist the atom too -the quarks included-, we cannot hope that we shall ever be able to see those particles or their subdivision, if of course, they exist, in a common way.

An experimental answer to the above question, whether a quark is a divisible or indivisible particle of matter, by direct observation of the quarks may be a very farfetched issue. At the moment, however, what can settle things is a theoretical answer. On the other hand, the theoretical answer that will be given will have both to be completely adjusted to the data existing today and to be quite convincing and understandable in order to clarify the answer to our question about the divisibility of the quark and not complicate it even more. In this way, I feel that the answer I describe constitutes an important step towards the solution of this problem.

In any case, the answer I describe does not only answer the question whether quarks are elementary particles, but leads us also at last to the real elementary units of matter, from which the other particles were formed. The "theory of the chain reaction" and the theory of the "unification of the fundamental forces and the physical theories", as well as the theory of the "formation of matter and antimatter" were based on the subdivisions of the quarks, will describe below.

THE ANSWER TO THE PREVIOUS QUESTION, WHETHER QUARKS ARE DIVISIBLE OR INDIVISIBLE PARTICLES AND THE SUGGESTION ABOUT THE NEW ELEMENTARY PARTICLES "POINTONS" AND "ANTIPOINTONS"

Let us try to form mentally a probable mechanism with which particles with a particular charge and particles with "exactly double charge" can be produced —as it happens with the quark "up" and the quark "down"—; we shall then discover that the simplest, most possible and reasonable way is to create a mechanism that will produce the smallest particles and the bigger ones will be produced by the doubling of the smallest particles that we have already produced. This is also the way in which nature, that always takes the simplest and most reasonable course of evolution, selected a similar mechanism for the formation of the quark up. Thus, we can assume that at least the quarks up are "divisible particles".

At this point I shall remind the readers the case of the discovery of the divisibility of the atom, when, around 1910, Joseph Thomson noted that there are particles, the "electrons", which have a mass much smaller than the mass of the atom of hydrogen. So the same question rises here again; could we have a similar case with the quark up, like the one we had with the atom and the electron, leading us to the indication that quark up, and perhaps all types of quarks –as I describe in the "theory of the chain reaction"– are, in their turn, divisible particles? Indeed, in this case of the quarks we have a clearer and more reasonable indication that they really are divisible particles.

Therefore, according to the above thinking, we are led to the both undisputable and self-evident theoretical conclusion without needing an experimental confirmation –at least at this stage of our research– that the division of matter does not end at the quarks but they too must be divisible particles in their turn, i.e., they must be formed by other smaller particles.

We shall extend this assumption, that quarks are divisible particles of matter too, to the other particles with a charge multiple of that of the quark or the charge of the suggested particles, pointons or antpointons, as I describe. For example, I would like to mention that, besides quarks, protons and neutrons, we should consider electrons divisible particles too, as their charge is a multiple of the charge of the quark.

However, our problem is neither solved nor ended with the discovery and the statement that quarks have subdivisions; but it remains, as we have to find out what these subdivisions are and study –at least theoretically– the probable mechanism by which quarks are consequently formed from these subdivisions.

In this case, then, I shall suggest a new, elementary, indivisible particle without *mass* or *dimensions*, with a charge equal to 1/3 of the charge of the proton, which I shall temporarily name it "pointon". Together with this particle I shall also suggest its respective antiparticle, the "antipointon", which has a charge exactly opposite and equal to 1/3 of the charge of the electron.

The selection of these tow particles, pointons and antipointons, was not made randomly, but it was made because these particles constitute the smallest subdivision of charge found until our days, as well as, as we shall see later, these particles fulfill all the requirements to be the smallest, indivisible, without mass and dimensions unit of charge and matter. At the same time, the selection of these particles will constitute the basis for the successful development of the "theory of the chain reaction", as I describe it in chapters six and seven hereafter.

With the selection of the pointons and antipointons, as we shall discuss in the "theory of the chain reaction" the elementary components of matter are now limited to the above two particle charges, which have neither mass nor dimensions and from which all the other particles of matter and antimatter were formed. Certainly, lots of other particles were also formed from the pointons and antipointons; however, as we are going to discuss in the "theory of the chain reaction", those did not need to play any specific, vital role in the formation of the material bodies and then in the creation of the Cosmos.

THE FUNDAMENTAL FORCES

In the previous sections of this chapter we described the structure and the behavior of the particles that constitute the foundation stones for the formation of the Universe. However, for the joining of these particles and the formation of matter and then the organization of the Universe, a series of other factors, the role of which was to mount, join together and arrange those foundation stones, was also necessary. These factors were the forces created among the particles and the masses of material bodies.

Until now, physics needed and distinguished four forces in order to describe the various physical phenomena of the Universe; due to their vital roles they were named fundamental forces. These four forces are: the "gravitational force", the "electromagnetic force", the "strong nuclear force" and the "weak nuclear force".

The first two forces, that is the gravitational force and the electromagnetic force are very well known by their applications, their behavior and their absolutely obvious results on our everyday life.

Especially, the "gravitational force", or the "force of gravity" as it is usu-ally called, is the force exerted between the masses of two material bodies. It has been studied by Newton, who formulated the law of the universal gravitation, which is expressed by the formula:

$$F = G \frac{M_1 \times M_2}{r^2} \tag{1}$$

Where $M1$ and $M2$ are the masses of the two material bodies, F is the attraction force developed between them r is the distance between them and G the gravitational constant among these masses.

The gravitational force is the force retaining masses on Earth, Earth on its orbit around the Sun and the stars and the galaxies in specific places within the Universe. It is the responsible force for the weight of material bodies. It is always an attraction force, with great range; however, for material bodies with conventional mass quantities it is a very weak force, with almost imperceptible power. Gravitation is one of the four fundamental forces that dominate and govern the Universe; in fact it is the most important and basic force in the megacosm and in the cosmos of infinity, whereas its effect on the microcosm is negligible or rather inexistent.

Due to the fact that the gravitational forces are very weak it is very difficult to study and investigate the nature and the medium carrying these forces. The results from the study and the nature of the gravitational forces, the medium carrying them and the investigation of their relation with the electromagnetic and the other fundamental forces are for the moment negative.

Matter, Elementary Particles and Fundamental Forces

The medium carrying these forces has not been found yet, which means that the cause for the attraction of masses has not been found. Recently the suggestion that the gravitational forces are due to the exchange of the hypothetical particles called gravitons and that it is formed by the exchange of those particles among material bodies prevailed.

Such a suggestion, however, that is, that the gravitational forces are due to the exchange of graviton particles among material bodies is completely hypothetical and arbitrary and has by no means been confirmed yet. The said suggestion that gravity is due to graviton particles is just a guess without any proof as the gravitons, as particles, have not been found yet. At the same time, however, as we are going to discuss later, the suggestion that gravitons indeed exist is completely contradictory in many cases to the real properties of gravity, as it functions and behaves within nature. Therefore, in general, we can assume that until today there is not the least confirmation proving the existence of gravitons and the way in which these particles are connected to matter and gravity.

Some of the most basic questions rising in the case of a hypothetical acceptance of gravitons as particles responsible for the gravitational force are:

a) What might the mechanism that formed the gravitons be?
b) What are the position and the relation of those particles with matter and in which way do they incorporate within the particles that form matter?
c) What might the mechanism that so harmoniously incorporated the gravitons within matter, without any failure or fault, be?
d) What should the graviton size be so that they are able to provide all those completely different mass sizes (in decimal scale) of the particles?
e) To what level can their division continue and to what point will the respective division of matter continue?
f) Why might gravitons "as mass carriers" exist wherever photons "which are charge carriers" exist and how is it possible that

gravitons and photons have exactly the same range and the same properties except power and polarity?

In the "theory of the chain reaction" all interactive particles, and therefore gravitons too, are replaced by the levels of the "electromagnetic interaction", as described analytically in the theory. In the "theory of the chain reaction" I also describe a property of the gravitation that has not been noticed until now, the following: when two atoms approach more than a defined limit, their attraction, instead of increasing, contrarily diminishes and in the end becomes zero and then it becomes repulsion. The theory names this property "dynamic equilibrium of the mass" in distinction to the property of the "dynamic equilibrium of the nucleus", which is accepted today by physics; a property valid for the "strong nuclear force".

The "dynamic equilibrium of mass", although as a property, has not been noticed until today, it plays an important role in the formation of matter, antimatter and then in the evolution of the whole Cosmos. In the third book of the trilogy of the creation, I shall describe also, another property of the dynamic equilibrium of mass that is so complex that cannot originates by particles and as we shall analyze in this case if assume to use particles, these particles should have extremely complicated properties, so complicates that it is impossible to be created by simple particles.

The second fundamental force, familiar also from our everyday experience, is the "electromagnetic force", the force that develops between elementary particles such as electrons and protons and material bodies with electrical charge. It is governed by Coulomb's law, which is expressed by the following formula (2).

$$F = K \frac{Q_1 \times Q_2}{r^2} \tag{2}$$

Matter, Elementary Particles and Fundamental Forces

In formula (2) Q1 and Q2 are two electrical charges, F is the attraction or repulsion force developed between them, r their distance and K the universal electrostatic constant.

It is an attractive force when exerted between heteronymous charges and repulsive when exerted between homonymous charges and it is much more powerful –about 10^{36} times– than the gravitational force.

The electromagnetic force makes electrons spin around the nucleus –as gravity makes Earth move around the Sun– and is the regulator of the chemical reactions in nature. We shall encounter this force in a lot of everyday electrical applications.

The attraction produced between the charges, at least as accepted until today by physics, is due to the exchange of particles of matter without mass, the photons, and originates from the mass difference –one millionth of a millionth of a gram– upon the formation of the atom, i.e., upon the bonding of the electron to the nucleus. In this way, the electron and the proton are heavier when studied separately than when studied as elements within the atom.

The other two fundamental forces are: the "strong nuclear force" and the "weak nuclear force" and relate only to the microcosm. Among those, the "strong nuclear force" is the one interacting with the quarks forming protons and neutrons and then, from protons and neutrons, the nuclei of the atoms. It has the property to join only specific combinations of quarks. It also joins several combinations of quarks and antiquarks forming a king of unstable particles named mesons. Those, in turn are immaterialized among themselves producing electrons and various other unstable particles. The production of mesons and electrons from quarks and antiquarks is one more confirmation that quarks and antiquarks are divisible particles.

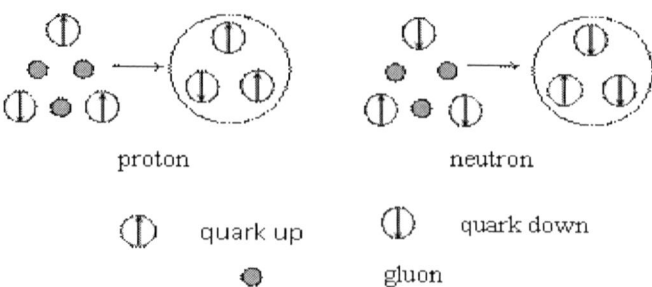

Figure 10. The formation of proton and neutron from quarks.

The particle named gluon is considered the carrier of the strong nuclear force. Both quarks and gluons have not been observed until now as individual free particles in nature. Figure 10 shows the combinations of quarks and the forces that interact between them in order to give us the protons and neutrons, which then join under the same forces –as it is accepted by current physics– for the formation of the nuclei.

In the very small subnuclear distances, the strong nuclear force is reversed and acts reversely hindering the subnuclear particles to approach in a distance smaller than a specific limit. This property was named "dynamic equilibrium of the nucleus".

The above property was established for the description and the reasoning of the case of coexistence of subnuclear particles in the nucleus of the atom without their dissolution –attractive effect– or their self-destruction when they approach closer than a certain limit –reversal of the attractive effect–. Physics has not yet provided any convincing explanation about the nature of the "dynamic equilibrium of the nucleus".

In the "Trilogy of the Creation", the behavior of the strong nuclear force within the nucleus can be understood based on the function of the levels of the electromagnetic interaction from which the strong nuclear force is formed. We are going to study these cases analytically, as stated above, in the second and third books of the trilogy. However, in general, in the "theory of the chain reaction", the "strong nuclear force" as well as the "gravitation" can be fully understood as main levels of the "electromagnetic interaction".

Finally, the fourth fundamental interaction concerns the "weak (nuclear) force", the force that produces radioactivity and is encountered in reactions of the radioactive fission of particles. It is exercised between particles of matter with spin 1/2, but it does not exist between particles – carriers of interaction with spin 0, 1, 2, such as gravitons and photons.

Carriers of this force are the particles W^+, W^- and Z^0, which appear in experiments with low energy levels, when their mass increases and their range is reduced. However, in higher energy levels, these particles behave as one particle and then, in even higher energy levels they behave just like photons. In any case, although the mechanism of this force has been fully understood, the causes that trigger this mechanism have not been understood yet.

Table 1. The properties and the carriers of the fundamental forces as accepted today by physics

	Exercised	Magnitude	Range	Carrier
Gravitational force	Between masses	10^{-38}	Infinite	Graviton
Electromagnetic Force	Between electrical charges	10^{-2}	Infinite	Photon
Strong Force	Between particles	10^0	$<10^{-15}$ cm	Gluon
Weak force	Between electrical charges	10^{-6}	$<10^{-17}$ cm.	W^+, W^-, Z^0

Table 1, above show the properties of the fundamental forces, as they behave and function within nature. The same table includes also the carriers of those forces. These data are the ones physics accepts until today.

Note: In chapters six and seven, where the "theory of the chain reaction" is described, we shall prove that the particles carrying the fundamental forces do not exist and that not only do they not exist but at the same time physics do not need these particles. In parallel, we shall prove that all the above forces and all their carrier particles are summarized in just one interaction, the "electromagnetic interaction", which takes place between the elementary particles pointons and antipointons.

So the "electromagnetic interaction" produces the attractive or repulsive force between the pointons and antipointons. This is the electromagnetic force accepted today by physics. The other fundamental forces except the "weak nuclear force", i.e., the "gravity" and the "strong nuclear force" result as primary levels of the "electromagnetic interaction", automatically by the formation of the respective particles, without the need of energy, high pressure or temperature change, as physics accepts. Specifically, as we are going to describe in the "theory of the chain reaction", together with the formation of the protons and neutrons the "strong nuclear force" is triggered, and with the formation of atoms the "gravitational force" is triggered too.

In the "theory of the chain reaction" we shall also see that all the other interactions, such as "magnetism", "light", "heat", "weak nuclear force" –which, according to the "theory of the chain reaction" is not considered a fundamental force–, etc., are proven to be secondary levels of the "electromagnetic interaction" and are created automatically depending on the way the subnuclear and subatomic particles move in the nuclei and in the atoms, respectively.

Chapter 5

THE COSMOLOGICAL THEORIES AND THE THEORY OF THE "BIG BANG"

> Ed. Hubble's observations led us to the thought that there has been a moment, the moment of the big explosion, as it is called, that the Universe was infinitely small and infinitely dense. At the same time that is at the moment of the big explosion, as the Universe was infinitely small it should also be infinitely warm.
>
> From Stephen Hawking's book. *A Brief History of Time*

THE COSMOLOGICAL THEORIES

The term cosmological theory qualifies a theory describing the creation, the evolution and the functioning of the Universe. Among the existing cosmological theories, the theory of the "stable condition" has prevailed for many centuries and until recently. According to the theory of the stable condition, the Universe had no beginning and has been always stable. And, as the theory assumed, new matter was continuously created in order to fulfill the voids between galaxies.

Nowadays, however, physicists, astronomers and cosmologists, support a new theory that replaces the previous one of the stable condition. Scientists named this new theory the theory of the "big explosion - big bang".

According to the theory of the "big bang", the whole Universe, at the beginning, was limited in a very small ball, smaller than an egg. This small ball contained infinite energy, infinite density and had infinite temperature. At a certain moment, a very rapid expansion of this ball started; so rapid that it seemed as if an enormous explosion was taking place. The Universe was created from this explosion, as assumed by the theory and as I describe analytically in the next section.

However, among all the cosmological theories that have been suggested until now, not one, the theory of the "big bang" included, has been able to provide a convincing explanation about the creation of the Universe. Mostly, the cosmological theories that have been suggested fail to describe the initial periods of the creation. Also, they cannot unify the individual prevailing theories, i.e., the theory of "relativity" and the theory of "quanta" in a one whole that will explain in a convincing and clear way how the events within the Universe began and evolved. At the same time, the existing theories cannot provide clear information with a certain continuity and compatibility with the actual condition and function of the Universe.

Therefore, under such circumstances, all the existing cosmological theories are disputable, are considered surpassed and nowadays physics seeks a new model that will describe the creation in a more convincing manner, a model that will then be able to unify all the individual theories, opinions and perceptions about the Universe.

All of the above constitute one of the reasons why, in the analysis of the various cosmological theories hereafter, I shall keep to the description of the theory of the "big bang" only, which is the most prevailing one today for the creation of the Universe, considering that there is no particular reason, to tire the readers, describing ideas that are no longer valid.

THE THEORY OF THE "BIG BANG"

So, according to the theory of the "big bang", 15 to 18 billion years ago, there was absolutely nothing in the Cosmos but a very small ball with infinite density, infinite energy and, consequently, infinite temperature; its size did not surpass that of a very small egg. This is the reason why scientists named this ball "cosmic egg". The whole actual Universe was concentrated within this small ball.

Absolute rest dominated everywhere. There was neither light nor dark, neither cold nor heat, there was no matter and the concepts of physical laws and physical phenomena were inexistent. The only thing that existed within the Cosmos at that time, as the theory assumes, was this small ball mentioned above. At a certain moment, the ball started to grow precipitously, so precipitously and so rapidly that an enormous explosion seemed to take place, so big that it is difficult even to imagine or describe it.

At exactly the same moment, always according to the theory of the "big bang", the ball started to cool down and simultaneously a series of events began; these events created the Universe as we see and feel it today and as describe below.

As the theory assumes, at the time the explosion of the ball started, time also began. Also at the same time, space started to be created. So, together with the beginning of the big explosion, we also have the beginning of time and the beginning of the creation of space.

(Here, with a small parenthesis in the description of the events of the theory of the "big bang", I remind the readers that in chapter one, we expressed several completely different perceptions about the beginning of time and the creation of space. We considered that time and space existed before the big explosion and in infinite quantities, but in a certain latent condition.

In particular, we claimed and proved that both time and space existed before the "big bang", as these concepts are completely independent from the various events, and therefore there was no reason at all for them not to exist even before the big bang, which we can describe as an individual event within space and time. According to this reasoning, an event can start and come to an end within a certain time and within a certain space, as for example the creation, the evolution and the end of a Universe; however, space and time did not start and will not end with the creation and the end of this Universe.

But all the above is enough about our perception of time and space and so, closing this short parenthesis, we would like to continue the description of the series of events that followed after the big explosion, as assumed today by the theory).

So, initially and within a very short, almost infinitesimal period of time after the moment the big explosion started and up to 10^{-43} seconds, a big ball of fire was formed, which was a mash consisting of a mixture of energy and unknown micro particles of matter and antimatter. The temperature within the ball dropped sharply, from infinite in the cosmic egg at the moment the big explosion started, to 10^{30}°C.

Before the specific time of 10^{-43} seconds, according to the theory, no physical laws and absolutely no physical principles applied and there is no theoretical idea that might describe this period of time. For this reason, science describes this period that is the first period, as the "miracle of the creation".

Then and again in the infinitesimal time period of 10^{-43} seconds to 10^{-32} seconds, particles of matter started to separate from energy and light radiation turned into matter. At that time various unknown particles started to form and the first interaction between those particles started to take place. The temperature dropped gradually from 10^{30}°C to 10^{27}°C. This second period, which is also not clear enough, was named the period of the "grand unification".

The third period lasted from 10^{-32} seconds to 10^{-10} seconds and its characteristic is the formation of the elementary particles; quarks and antiquarks, electrons and antielectrons. However, with the quarks, the antiquarks, the electrons and the antielectrons –always according to the theory– another type of finer particle was formed, the gluons, which united the quarks and created the components of the nuclei, as we shall see next.

The Cosmological Theories and the Theory of the "Big Bang"

The fourth period followed, during which the antiquarks and the antielectrons disappeared gradually and from the quarks, protons and neutrons were formed. Meanwhile, the temperature had dropped to 10^{10}°C. This period lasted from 10^{-10} seconds to $10^0 = 1$ second. Certainly, during the above periods, lots of other particles were formed, that did not assume a particular, active role in the subsequent processes.

The fifth period lasted from 1 second to 3 seconds after the beginning of the big explosion. During this period, the nuclei of the atoms of hydrogen, helium, lithium and deuterium were formed. The temperature had dropped to 10^9°C and then, slowly, appropriate conditions for the formation of the atoms and matter started to develop.

Then we had a much longer period, the sixth one, which lasted from the third minute after the big explosion until 300,000 years. During this period the quarks and the gluons disappeared and the weak nuclear force developed, which dominated matter; the electromagnetic force; and then the "gravitation force". Slowly, the first atoms of hydrogen and helium started to form and the temperature dropped now to 3,000°C.

Three periods followed, that is, the seventh period, which lasted about 1,000 million years and during which matter separated from radiation; the eighth period, which lasted about 14,000 million years and during which matter accumulated creating the Quasars, celestial bodies with great brightness and strong radiation, the stars and the protogalaxies; and finally the ninth period, which started about 5,000 million years ago. During the ninth period, the Galaxies and the Solar systems were formed, the Planets and the Earth, complex molecules and living matter. The temperature of the Universe dropped to 3°K. This period lasts until now.

At the end of this chapter I attach Table 2, where the timing of the evolution is shown, with the periods of the creation of the Universe and the characteristics of these periods as the theory describes them.

WHY THE SCIENTISTS BELIEVE WHAT THE THEORY OF THE "BIG BANG" SAY; THE UNANSWERED QUESTIONS OF THE "BIG BANG" THEORY

Many scientists believe today the theoretical model expressed by the theory of the "big bang". The most sound reason leading scientists to believe in this theory is the observation of the astronomer and physicist Hubble that the galaxies do not have steady positions within the Universe but move with very high speeds, at times reaching the speed of light, and also that they draw away from one another.

This increase in the distance between the galaxies and its rate led the scientists believing in the "big bang" to conclude that the Universe started from a central point from which the big explosion began too. This central point is also indicated by the respective equations of the theory of general relativity, which foresee the convergence of all the components of space-time in a "mathematical abnormality" that, according to scientists is identified with zero point that is the beginning of the big explosion.

However, despite the predictions and the insistence of the scientists, Einstein explicitly refused to accept the theory of the "big bang", mostly the part of the theory which relates to the drawing away of the Galaxies and the expansion of the Universe. Indeed, sensing that some would try to involve his name in a theory which he rejected, he introduced the equations of the theory of general relativity, the well known geometrical constant, in order to preclude the probability of the explosion of the Universe by their new formulation. Later, however, when the expansion of the Universe was definitely confirmed, Einstein claimed that the introduction of this constant was a mistake of his and for this reason he then retracted it.

However, besides the expansion of the Universe, scientists invoke also two other serious experimental reasons which stand for the establishment of the theory of the 'big bang". These reasons are:

The indication, concerning the quantities of hydrogen, helium and the other elements, exist in the Universe. The various measurements have shown that the composition of the Universe is 75% hydrogen, 23% helium and only the remaining 2% corresponds to other elements, such as oxygen, nitrogen, carbon, etc., which are abundant on Earth and very familiar and useful to humans.

So, based on the laws of nuclear physics, we can calculate the percentage of hydrogen and helium that should have been produced in the Universe by the processes of the big explosion. The theoretical calculation of these percentages is in absolute harmony with the percentages of hydrogen and helium found today in the Universe. Therefore, these data provide even indirectly an experimental support to the theoretical model of the "big bang".

The second indication invoking the "big bang" refers to a theoretical calculation by G. Gamow, a physicist, according to which, when the Universe was at the age of 700,000 years, a light radiation was produced, travelling today all over the Universe. So, according to G. Gamow's calculations, today this radiation should have reached the temperature of 3°K.

G. Gamow's calculation remained almost unnoticed until 1965, when A. Penzias and R. Wilson made a historical discovery, accidentally as it often happens with discoveries in physics.

A. Penzias and R. Wilson, while trying to trace the signals of an artificial satellite, observed a steady and inexplicable radiation on their receiver. It was found that this radiation is the light that was released by matter billions of years ago, that is 700,000 years after the big explosion, and that the cold light is the one corresponding to this radiation, which today has indeed the temperature of 3°K. Certainly this radiation is uniform towards all directions of the Universe. This is something that seriously troubles scientists, as this uniformity cannot be explained by the data of the "big bang".

The observation of the drawing away of the Galaxies and the indications of the percentages of the elements that exist in the Universe and of the light radiation matter emitted when the Universe was at the age of 700,000 years established the "big bang" as the most predominant theory of creation. This of course happened with the supposition that the theory will provide clear answers to several other questions too that have not yet been answered.

The most important of these questions are:

Where was the "cosmic egg" found, with the infinite energy, the infinite density and the infinite temperature, from the explosion of which the whole Universe was created, and how was it formed?

Where and how were this infinite energy, infinite density and infinite temperature contained within the "cosmic egg" manifested?

How was the energy of the cosmic egg transformed into matter? Specifically, how was this energy transformed and how were the quarks and then the other particles of matter formed?

Are the quarks elementary particles, i.e., are they indivisible particles or does the division of matter proceed even further than the quarks?

What mechanism created the unified interaction that contained the four fundamental forces, which developed gradually and led to the evolution of the Universe?

What are the particles that were the carriers of the above forces and how were they formed; and how are those particles incorporated in matter?

How were the nuclei of the atoms formed?

What are the causes that created gravity?

What is the particle called graviton and where is it found?

What mechanism acts in order to change the sign of the "strong nuclear force" and render it from attractive to repulsive when the particles that carry it approach closer than a particular limit?

Recently one more basic question rose and it was added to the list of questions expecting an answer. It was noted that the galaxies in our Universe do not move with a steady speed but their movement is at the same time accelerated. In this case, "physics" as a science and the "big bang" as a theory, seek a convincing answer about the cause for this acceleration of the galaxies.

However, about two centuries have passed since the establishment of the theory of the "big bang" and the above questions remain still unanswered. This causes a great deal of insecurity in what concerns our knowledge about the evolution of the events of creation. In order to realize the magnitude of this insecurity we should proceed to the following very simple intellectual thought:

We described above the questions that rose from the establishment of the theory of the "big bang", which as readers of the present work we read as a simple text without paying them much attention or attributing particular importance to them. The thought I suggest is to reread the questions paying more attention and try to provide the appropriate answers. What might the result be?! Shall we be able to answer at least one question?! In any case, our surprise will be even greater when we think further about several other questions, such as what laws applied during the first moments of the creation, what happened to antimatter, etc. questions we expressed in previous sections.

At this point, leaving behind much vagueness and a lot of questions about the creation, conclude this chapter about the cosmological theories. At the same time, conclude also our short intervention concerning the description of the evolution and what happens in the Universe according to the ideas of physics that have been established until now.

Table 2. The timing of the creation of the Universe, according to the "big bang" theory

1st period	2nd period	3rd period	4th period	5th period
the miracle of the creation	the era of the great unification	the era of quarks and antiquarks	the formation of protons and neutrons the disappearance of quarks	the formation of nuclei of Hydrogen and Helium
temperature ∞-10^{30} Duration 0-10^{-43}sec	temperature 10^{30}-10^{27} Duration 10^{-43}-10^{-32}sec	temperature 10^{27}-10^{15} duration 10^{-32}-10^{-10}sec	temperature 10^{15}-10^{10} duration 10^{-10}-10^{0}sec	temperature 10^{10}-10^{0} duration 1sec-3min

Table 2. (Continued)

6th period	7th period	8th period	9th period
the joining of matter and antimatter	the separation of matter and antimatter	the formulation of Quasars, Stars, and Proto-galaxies	the formulation of the Galaxies and Solar systems
temperature 10^0-3.10^3 duration 3m-3.10^5 years	temperature 3.10^3-18^0 K duration 3.10^5-10^9 years	temperature 18^0 K-3^0 K duration 10^9-15.10^9 years	temperature 3^0 K duration 15×10^9 years

Automatically, however, in the chapters that follow, describing the "theory of the chain reaction", the writer's role is reversed and from judging, as happened until now, he will be judged. This is why I tried to be as clear, simple, explicit and understandable as possible in the description of the "theory of the chain reaction"; in order to give the reader the chance to understand the theory and then, after having understood it, to be able to proceed to the adequate criticism.

Chapter 6

THE THEORY OF THE CHAIN REACTION: A NEW ATTEMPT TO EXPLAIN THE CREATION AND THE OPERATION OF THE COSMOS

In the chapter about the "unification of the fundamental forces and physical theories", when describing the process and the definition of the concept of unification, I mention that we can present the total of the physical theories as a tree with multiple branches that represent the various physical theories. Unfortunately, however, the root of the tree, some parts of the trunk and many of its branches were lost during the infinite years of the past. Science is currently making efforts to discover these lost parts in order to form a complete, clear, convincing and understandable picture of this tree.

Under these circumstances, I believe that the "theory of the chain reaction", which I describe in this chapter and complete it in the next, will decisively contribute in the efforts of science for the completion and understanding of this picture of the tree.

The writer's point of view

GENERAL COMMENTS

I read, in too many textbooks, that in order for a theory to be accepted, it must meet several basic criteria. In all cases, scientists accept that, firstly, a theory has to agree with the already existing facts; secondly, it has to be clear and understandable, using the least possible arbitrary axioms and assumptions; and, thirdly, it must be self-checked and provide predictions, which allow to check, its correctness and its credibility.

Taking into account the above criteria and agreeing with them, I noted that the various cosmological theories I studied at times, did not meet not even a single one of them. And if one of those theories met one of the criteria, it did not meet the others.

For instance, the theory of the "big bang", as mentioned in the previous chapter, is based on a lot of arbitrary axioms and leaves many unanswered questions. At the same time, I recalled about the "theories of relativity" that when a student asked the famous astronomer and professor, Arthur Eddington, an authority in the field of these theories, if indeed only three people in the world understand them, Arthur remained hesitant for a while. When his interlocutor asked him what he was thinking, he answered: "I am trying to understand who the other two scientists that understand the theories are!"

So, on what rationale can we suppose that under such conditions, the "big bang" and the "theories of relativity" meet the above criteria, since they do not meet the second one, "that the theories should be clear, understandable and using the least possible arbitrary axioms and assumptions"?

However, I have to admit that I made a lot of efforts to find something more worthy, but I concluded that none of all the cosmological theories I had read could explain with at least elementary credibility and clarity, how the Universe was formed and especially, describe the initial stages of its creation.

So, having the above in mind and evaluating some data I had collected and having considered these data most credible from the existing ones, I decided to write the "theory of the chain reaction", which I feel meets, more than the others, the criteria a theory needs in order to be accepted.

In the following sections I describe the processes that, according to the "theory of the chain reaction", created, the Universe, the Universes, the Anti-universes and the whole Cosmos, as we know and as we perceive it.

How Were the Pointons and the Antipointons Formed? The "First Stage of the Creation"

At the end of the first chapter we left a "Cosmos", consisting of an absolutely empty space, of infinite size. Nothing there was inside this space, neither cold nor heat, neither light nor dark, neither black nor white, no matter, no energy; the absolute zero prevailed everywhere. And not only were all the above inexistent, but they had not yet been manifested, not even as concepts. The only thing that existed in that absolutely empty Cosmos, except the infinite "space", was the infinite "time" also. Both "space" and "time" was in a latent state as there were no events to be used for a comparison, a measurement or an observation or even a perception of these concepts.

This state may have lasted only several moments or even many millions of years. However, we are not interested in how long the Cosmos had the above form, as we are not able to measure this period because there were no events then and there was no need for a measurement under the above circumstances and conditions, as a measurement like that would have no value or sense at all. We can characterize this state as the state that constituted the furthest time boundaries of the past.

Under such conditions, the simplest move or, in other words, the simplest event that might happen in this absolutely empty Cosmos, –an event that would modify the state of the absolute rest, but would also evolve, be adjusted, harmonize with and be compared to the current real conditions in our Universe–, was the creation of an abnormality –and more reasonably–, the utmost small and imperceptible abnormality that might exist. The "theory of the chain reaction" accepts as such abnormality, the creation of an elementary particle, the *"pointon"* the properties of which have been already described in chapter four. We shall consider the formation of the pointon the first event that took place in the Cosmos and at the same time this event signaled the beginning of the creation. At the beginning of the creation, the only characteristic property of the pointon, that could be distinguished, was the elementary electromagnetic radiation it emitted, which corresponded to the electromagnetic radiation as this is accepted today by science. The power of the electromagnetic radiation of the particle pointon, as described, is equal to *1/3* of the power of the electromagnetic radiation of the proton.

After the formation of the pointon, the electromagnetic radiation started to disperse inside the absolute void at a very high speed, the range of which corresponded to the speed of the propagation of the electromagnetic waves. The equilibrium and the rest of the absolute void had at last been disturbed. When the electromagnetic radiation reached a certain point in the space, the absolute void ceased to exist. We can consider the speed of the propagation of the electromagnetic radiation as the highest speed in the Cosmos, given that the speeds of the other events were gradually reduced as the Cosmos became increasingly more complex. Taking into account the above speed, we can calculate that the electromagnetic radiation, in a period of just one second, had created a vast spherical field with radius $R = txc = 1sec \times 300,000,000 \ m/sec = 300,000,000m$, where *"R"* depicts the radius of the spherical field created, *"t"* the time of one second and *"c"* the speed of the propagation of the electromagnetic radiation.

In order to react, cover and counterbalance the abnormality caused by the creation of the initial elementary particle pointon, the elementary particles, *"antipointons"* be formed, which were particles with exactly opposite properties than those of the initial pointon. The antipointons, in their turn, formed new fields of radiation opposite to the field created by the initial pointon. The formation of the antipointons was again considered a new abnormality and so new pointons were created for its counterbalance. In this way, a chain reaction started, producing pointons and antipointons, the rate of which, although enormous compared to the current rates of the evolution, increased in a geometric progression. This chain reaction, produced pointons and antipointons continues till today at the boundaries of the Universes and the Antiuniverses and we might assume that this rate will persist infinitely if there is no other abnormality that would set it back.

The pointons and antipointons created, as already described, had neither mass nor dimensions and their total energy was zero. This is the reason why the theory considered the pointons and antipointons point charges. Under the effect of the electromagnetic radiation, the heteronymous particles formed were attracted and the homonymous ones repulsed. The attraction aimed to join pointons and antipointons so that they would then counteract each other and end in their natural sum and their natural total energy, which was zero. On the contrary, the repulsion aimed to protect the particles formed.

At this point the question about the connecting substance that made pointons perceives the attraction of antipointons rises. Here, the "theory of the chain reaction" accepts that the medium that led the pointons to undergo the attraction of antipointons was a simple interaction –that we have already named electromagnetic interaction–, which did not need any interference by an intermediate connective material that would cause it.

We consider such a case as a fact, as the "theory of the chain reaction" accepts, that from one point and previously no material that would interfere between the pointons and the antipointons was necessary or could exist, given that the Cosmos started from nothing. On the contrary, the "theory of the chain reaction" accepts that the more and more the absolute empty space between these particles was the more perceptible was their interaction.

In commenting on the rationale of the theory, that no connective material existed among the pointons and the antipointons, we can say that, probably, it might seem that the theory introduces a hypothesis based on nothing, something like an arbitrary axiom. However, if we examine more closely the meaning of this hypothesis, we shall note that even in the case that we accept the existence of a substance between the pointons and the antipointons, this has nothing to do with the evolution and the course of the "theory of the chain reaction", as even if there is an intermediate material between the pointons and the antipointons, it was not necessary to play a specific role in the development of the theory. So, the study of the existence of an intermediate material among pointons and antipointons is a totally secondary issue for the theory and we shall leave it issue to be investigated by those insisting or who will benefit of such an idea.

All the above developments took place without consumption of energy; anyway, according to the theory there was no energy and the aforementioned developments were based on mechanisms that we have not been yet able to describe very clearly. This is the reason why the "theory of the chain re-action" invokes the contribution and assistance of an "Upper Unknown Force", of which we are not aware.

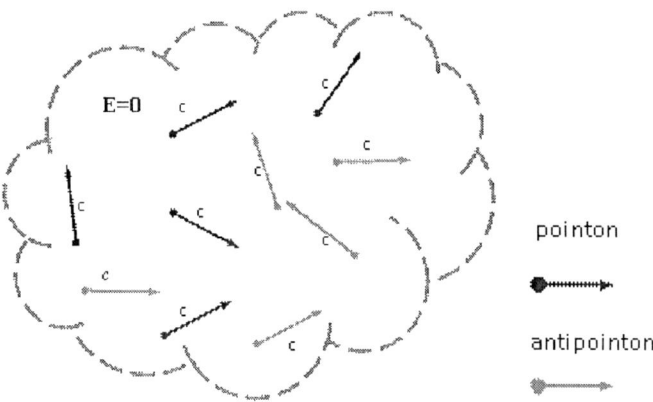

Figure 11. The Cosmos in the form of a set of pointons and antipointons.

However, with the beginning of the creation of the pointons and antipointons, we assume that the intervention of this "Upper Force" we have just mentioned stopped. From then on, until now, it left things develop by themselves, as we shall see; but under the "surveillance" of the "concepts" it had already created, which described in the first chapter and which had been kept hidden in order to be manifested and presented gradually, in parallel to the development of the events.

The only possible intervention of the "Upper Force" after the formation of the first elementary particles, pointons and antipointons, until now is when after the creation of a fairly large amount of matter, the beginning of the creation of the first elements of "Life" took place. We can say that this was the second and the last abnormality. However, the investigation of this issue is the object of a completely separate study beyond the scope and the boundaries of the "theory of the chain reaction".

The total of the pointons and antipointons formed behaved as a very unstable gas with zero energy and zero mass, as we have already described and as depicted in Figure 11 on the previous page. This was the whole Cosmos at that moment.

How the Laws of Nature Worked in the Early Moments of the Creation: The Movement of the Elementary Particles Pointons and Antipointons

As described in the previous section, the pointons and antipointons were entities, particle-charges, without mass or dimensions. Among these particles, an interaction took place, the "electromagnetic interaction" – which corresponded to the electromagnetic radiation–. This interaction, as we have already described, turned into attraction between the heteronymous particles or repulsion between the homonymous ones. The speed of the propagation of the electromagnetic interaction in the absolute void was equal to the speed of the propagation of the electromagnetic radiation. According to the theory of the chain reaction, this speed is not only the highest speed of propagation among all the other interactions – gravity, light, heat, etc.–, but is also a universal speed not just in our Universe, but in the whole Cosmos.

However, with the establishment of the elementary particle-charges, pointons and antipointsons, which were particles with neither mass nor dimensions and were moving at a finite speed, three basic questions concerning the laws that applied in their movement, their development and behavior rose; especially:

1) How is it possible to have particles, only charges, without mass or dimensions, moving at finite speed?
2) What happened at the initial stage of the creation; were the natural laws applied or not? And
3) How did particles with mass and dimensions result, from the elementary particles pointons and antipointons, as it actually happens inside the Universe?

In what concerns the first question, until now physics accepts that the movement of all material bodies –from the movement of the elementary particles to the movement of Stars, Solar systems, Galaxies and the Universe–, is described by the law of the Universal attraction and the three laws of the motion of the material bodies, which are, the law of inertia, – first law of motion–, the law of the acceleration, –second law– and the law of the action and reaction –third law–. These laws were formulated by Newton, are named after him and describe the dynamics and the motion of the material bodies.

If we try to apply the above Newton's laws to the motion of the elementary particles, pointons and antipointons, we shall note that these laws are inapplicable in the motion of particles without mass, as they laws have been expressed only for material bodies. For example, the application of the second law of motion "$F = m\gamma$" should result for the particles, pointons and antipointons, infinite acceleration, $\gamma = F/m = F/0 = \infty$ and, consequently, infinite speed, $v = v_0 + \gamma t = 0 + \infty t = \infty$, even under the effect of a minimal force. However, this is incompatible with the real conditions, as in the case of the infinite speed of the particles, pointons and antipointons, we would have unpredictable results that could not describe according to the existing natural laws. At the same time, these results would not be able to conform to the already existing conditions in nature. So, the second very serious and basic question rises. What happened during the initial stage of the creation; did the natural laws apply or not; and if they applied, how did this happen?

At this point, the existing cosmological theories, as they were not able to provide a complete description of the initial stage of the creation cannot give a positive answer to the above question and accept that the natural laws did not apply at the beginning of the creation. Contrary to the existing theories, the theory of the chain reaction starts with the axiomatic for the theory condition that the natural laws applied, from the first moment of the creation, either in their actual form or differentiated and adjusted in the conditions that existed at the beginning of the creation. In this case, the theory of the chain reaction considers that it is a big mistake to conclude that there were no laws that applied at the initial stage just because we are not aware of them. So, the theory of the chain reaction accepts and at the same time proves by its formulation, that anything happening in nature has always a cause, as well as a law guiding it. Thus, the events themselves that occurred during the initial stage include the natural laws that created them; we only have to discover those laws.

Reinforcing the above views, the theory of the chain reaction does not only accept that at the beginning of the creation all the natural laws applied, but, at the same time, it defines the laws that applied and is formulated based on those laws. Until now it is considered that in nature the forces are exerted on material bodies only. So the validity and the completion of the law of the universal attraction and the three laws of motion of the material bodies have been studied just for material bodies. However, with the establishment of the particles, pointons and antipointons, as charge entities without mass, the existing laws of motion of the material bodies become incomplete and inapplicable in the case of pointons and antipointons.

At this point, the theory of the chain reaction, instead of considering that due to the fact that the laws of motion of the material bodies cannot be applied on the elementary particles, pointons and antipointons; there are no laws governing their motion; on the contrary, based on its indisputable basis, extended the laws of motion of the material bodies and established the respective laws that apply to the new particles as well. So, the theory, instead of considering that the pointons and antipointons moved, functioned and evolved without any natural laws, selected the more reasonable possibility to examine what might apply in this case. The answer here was simple and given with the extension of the application of the laws of motion so that they will include the motion of particles without mass too. In this particular case, then, the theory completes the laws of motion of the material bodies and formulates them as follows:

Newton's three laws of motion, i.e., the law of inertia, the law of acceleration and the law of action and reaction, do not apply only when forces are exerted on material bodies, as it has been established and believed until now, but apply also for any entity on which forces are exerted. This means that if a force is exerted on an entity without mass, this entity will not necessarily move at infinite speed. At this point, nature's intervention, overlooking the existing laws, leads us by itself to the correct conclusion.

And, whereas according to the second law of motion, $F = m\gamma$, particles with no mass $m = 0$, such as the pointons and antipointons, should have infinite acceleration $\gamma = \infty$, resulting in infinite speed $\upsilon = \infty$ too, nature, ignoring the established laws, selected by itself the correct result providing those particles with inertia that causes finite acceleration and finite speed. At this point, the "theory of the chain reaction", based on data science possessed today, estimated also the speed of the movement of the pointons and antipointons, which is proportional to the speed of the propagation of the electromagnetic waves. It also estimated their acceleration, which is approximately 10^{38} meters per second. More on the extension of the above three laws will be provided in the work about the formation of matter and antimatter. In what concerns the law of the Universal attraction, this law did not necessarily apply during the initial stage of the creation, as no masses existed then.

However, at this stage of our study, we can accept that pointons and antipointons obey the same natural laws as the laws of the material bodies, if we take into account that the existing inertia of the pointons and antipointons gives them finite speed. We might say that this inertia is the sperm that exists in these particles, from which the rotating orbits around their opposite particles occur and form the other particles of matter, i.e., the quarks, electrons, protons, neutrons, nuclei, and atoms and then the masses and the material bodies. In this way, the third question we posed; how particles with mass and dimensions, resulted from the elementary particles, pointons and antipointons, which were just charges and had neither mass nor dimensions, is also answered.

Concluding, we may say that during the initial stage of the creation we had the particles, pointons and antipointons –simple charges, with neither mass, nor dimensions–, the electromagnetic interaction, –which is the same as the electromagnetic radiation, as this is accepted today by physics–, between these particles and the chain reaction producing pointons and antipointons. The electromagnetic interaction, together with the inertia of the pointons and antipointons, based on the absolute application of the natural laws, produced the rotating orbits of the pointons and antipointons. And as we shall describe, these rotating orbits resulted in, the other subnuclear particles, i.e., quarks, electrons, protons and neutrons and then the strong nuclear and the gravitational forces, from which the nuclei, the atoms, their antiparticles, the masses and the material bodies were formed.

THE EXPERIMENTS AND THE FATE OF ETHER

In the second section of this chapter, in the description of the first stage of the creation, the theory of the chain reaction assumes that the pointons sustained an attraction force from the antipointons as a simple interaction, without interference of a connective substance as a medium for this interaction. On the contrary, the theory considers that the cleaner the void between the above particles, the more perceptible their interaction.

A similar issue about the interactions of the elementary particles had been studied by physicists during the two previous centuries. The problem was posed together with the way of propagation of light and the electromagnetic waves. Knowing that the waves, in general, are propagated through a material medium, scientists believed in the necessity of the existence of a connective material that was considered to fill the whole Universe. They named this material for the transmission of the waves "ether".

However, the acceptance of the existence of a certain medium as a vehicle for the transmission of the waves resulted in the corresponding questions. For instance, as it was known that electromagnetic waves are transversal waves and as transversal waves, are propagated only through solid bodies, ether should behave as a solid body. In this case, can you imagine then, a Universe full of ether that would behave as a solid body, without any voids and without anything else?

Now, in what concerns the propagation of different waves, such as those of light, heat, sound waves, etc., many eminent scientists, like Cauchy, Stokes, Thomson and Plank considered that there were various ethers, with various properties, for each kind of transmission. Now, how could all those kinds of ethers coexist? It was another big question.

But, due to the fact that ether as a medium for the transmission of high speed transversal waves should be solid and rigid, new questions rose again such as, how could the Earth move in an unnatural environment like that?

A race for the detection of the ether and the study of its properties started so that these questions would be answered. Until the middle of the previous –twentieth– century, more than fifteen experiments had been performed for the detection of the ether and there are scientists who still insist in its existence. Among the experiments performed was the well known Michelson-Morley experiment, which was conducted in 1887 and was based on the separation of a light beam into two moving perpendicularly to each other. Then the two beams joined again. In this case, when the beams joined again they should have a phase difference due to the Earth's motion in the ether. However, all the experiments performed, the Michelson-Morley experiment included, although they used the most perfect instruments, were not able to detect a difference that would suggest the existence of ether.

Initially, this was attributed to the lack of precision of the experiments. But, as in most cases the researchers were very meticulous in the issue of precision, if ether existed the experiments should have yielded results; so the scientists were forced to admit that there is no ether and that the waves are propagated through the absolute void. So, instead of proving its existence, the experiments about ether essentially proved its inexistence.

During the course of my work, when I was writing the section about the first stage of the creation, I had not paid special attention to how physics had treated the subject of ether, as it was not necessary for the theory of the chain reaction to provide a rationale to something that according to the theory did not exist. However, I had written some paragraphs that justified the lack of such a medium during the interaction of pointons and antipointons that might be unnecessary. But then, when I studied about the experiments for ether, I felt very satisfied because I realized that the theory of the chain reaction, without aiming to it, had predicted a fact that science had already experimentally proven. So we may say that the theory of the chain reaction proves theoretically too the inexistence of ether. On the contrary, the experimental confirmation of the inexistence of ether by physics is a partial verification of the agreement of the theory of the chain reaction with the current data.

THE FORMATION OF THE QUARKS, ANTIQUARKS ELECTRONS AND ANTIELECTRONS THE "SECOND STAGE OF THE CREATION"

At the end of the first stage of the creation, we left a Cosmos consisting of an absolutely empty space, of infinite size, where at a certain spot, of this space, a cloud of elementary particles, pointons and antipointons, had been created; these particles were moving at a speed corresponding to that of the electromagnetic radiation. When pointons and antipointons approached, the following occurred:

a) Pointons collided with antipointons –or/and vice versa– due to the attraction that was exerted on them, resulting in their self-destruction, as shown in Figure 12.

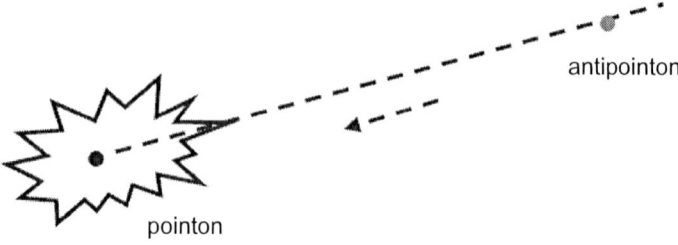

Figure 12. Collision of a pointon and an antipointon.

b) The second thing that could happen is that the antipointons would pass far from the pointons –or vice versa– and nothing would occur; antipointons and pointons would simply draw away from each other and continue their course among the other particles, as shown in Figure 13 and

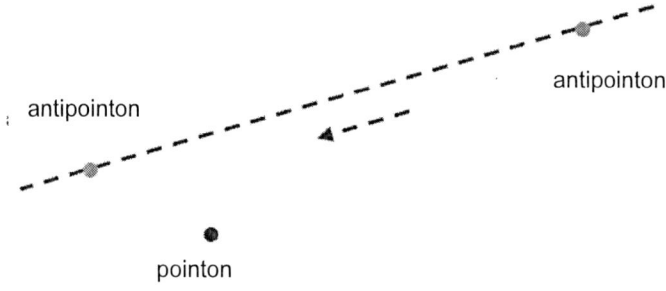

Figure 13. Case in which an antipointon is drawn away from the pointon.

c) The antipointons –or/and vice versa– would pass from such a distance away from the pointons that none of the two previous cases would apply, but the antipointons would be enclosed by pointons and enter in rotating orbits. In this way they formed new particles, as shown in Figure 14, in the following page.

The Theory of the Chain Reaction

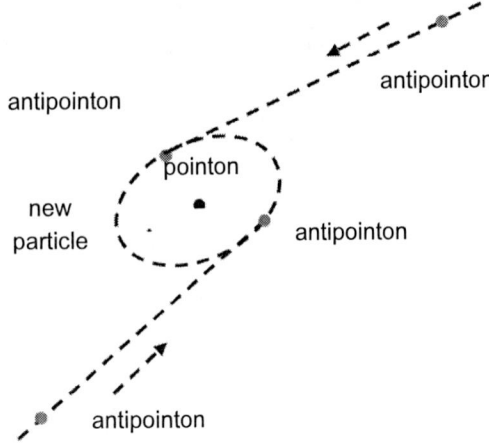

Figure 14. The formation of new particles by the particles pointons and antipointons.

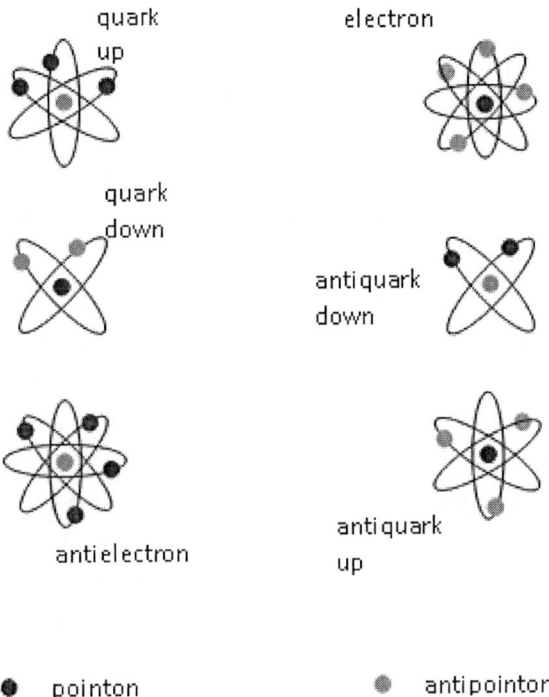

Figure 15. The combinations of pointons and antipointons that formed the basic particles of matter and antimatter.

After the above discussion, the following simple but very basic question arises: what were the particles formed from the pointons and antipointons?

The answer to this question is depicted in Figure 15, which presents all possible combinations with a total of up to five particles that might result from pointons and antipointons. Just one particle is found at the center of each combination, as, if there were more than one and heteronymous, they would have already been self-destroyed, and if they were homonymous, they would be repulsed and therefore they would not be able to be together at the center of the combination.

We shall be very surprised to notice in Figure 15 on the previous page, that all the combinations that occur between pointons and antipointons form the basic elementary particles of which matter and antimatter consist, i.e., the quarks up and quarks down, the electrons, their antiparticles and certain unknown neutral particles that might have a relation to neutrons.

The particles formed were charged particles as well. However, they possessed two additional characteristic properties that pointons and antipointons did not have. These properties were the *mass*, which was manifested as a very small residue of the sum of the attractive and repulsive forces exerted due to the interactions of pointons and antipointons, and the *dimensions*.

The residue of this force of the charges was created when the charged particles, pointons and antipointons, entered in rotating orbits around their opposite particles. Thus, instead of complete counterbalance of the forces exerted between the charges of the pointons and antipointons when these entered in rotating orbits, a very small residue remained, which at the first stages of the creation, remained hidden as the interactions due to the forces of the charges prevailed it by far. This residue started to be considerable when the formation of large concentrations of particles began. We shall name this stage of the formation of quarks, electrons and their antiparticles the "second stage of the creation".

The particles created were once more charged particles of matter and anti-matter of primary form. The fact that the particles formed were charged is the reason why they could not exist free in nature. So, at a subsequent step, they continued their reactions with the creation of a new series of particles, protons and neutrons which are now familiar to all. Then, protons and neutrons created the nuclei that, with the electrons, formed the atoms as are described in the next sections of this chapter.

THE FORMATION OF THE PROTONS AND NEUTRONS, THE "THIRD STAGE OF THE CREATION"

So the quarks, in turn, started to join, in their known now combinations, the most basic of which are those that created the protons and neutrons; that means: Two quarks up and one quark down formed a proton. And two quarks down and one quark up formed a neutron. The way in which, according to the "theory of the chain reaction" the above combinations took place is shown in Figure 16.

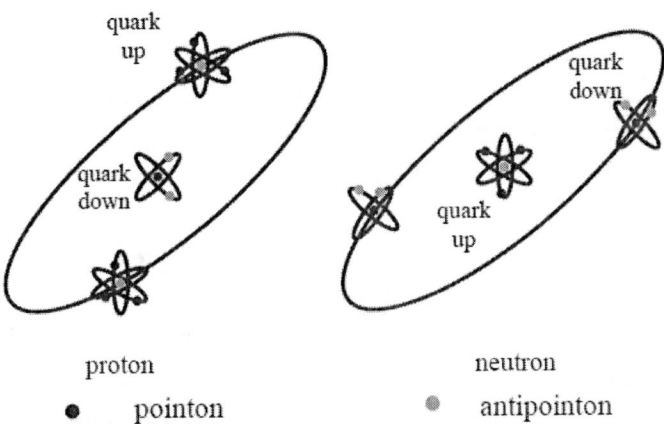

Figure 16. The suggested indicative structures of proton and neutron.

In this case, the quarks, –the speed of which had been reduced at the level of the speed of the electrons, measured when the latter spin around the nuclei of the atoms–, moved in rotating orbits, similar to those of the pointons and antipointons around their opposite particles, as shown in Figure 16. The figure depicts also the suggested indicative structures of the particles, protons and neutrons, as those could be formed after the suggestion about the establishment of the rotating orbits and rotating motion of particles inside the nucleus. We shall name this stage of the formation of the protons and neutrons the "third stage of the creation". Certainly, it is self-evident that in this stage, at the same time with the formation of protons and neutrons, their respective antiparticles were also formed in approximately equal quantities.

THE FORMATION OF THE HELIUM NUCLEI, THE "FOURTH STAGE OF THE CREATION"

With the creation of the protons and neutrons, a force was created between them the *"strong nuclear force"*. The strong nuclear force joined the protons and neutrons and created the nuclei of Helium, which consist of two protons and two neutrons. Together with the nuclei of Helium, created and their antiparticles

Of course, the nuclei of Hydrogen had already been created, without needing anything more because they are formed of simple protons. The nuclei of several other atoms were also formed, but in very small quantities, so in this stage, they did not play any specific role in the further evolution.

At the same time as the strong nuclear force created, the *"dynamic equilibrium of the nucleus"*, –which is the property that makes protons and neutrons to be repulsed when they come closer than a certain distance–, was also formed, This property contributes to the coexistence of the protons and neutrons in the nucleus and prevents them in their mutual destruction.

The mechanism, resulting in the strong nuclear force, the dynamic equilibrium of the nucleus and the nuclei, will be discussed and described in detail, in the third book of the "trilogy of the creation", were analyzes the theory of the creation of matter and antimatter

THE FORMATION OF THE HYDROGEN AND HELIUM ATOMS AND ANTIATOMS, THE "FIFTH STAGE OF THE CREATION"

After the formation of the protons, neutrons and the nuclei of the atoms of Helium and their antiparticles, the Cosmos consisted of the infinite void space, the "concepts" that started to emerge slowly, slowly and at a certain region of this infinite void space the elementary particles that had been formed until that moment, concentrated all together. These particles were the pointons, the quarks, the protons, the neutrons, the electrons and the nuclei of the atoms of Helium; together, the respective antiparticles had been formed and coexisted in the same space, in about equal quantities. These particles and antiparticles were reproduced at a rate higher than that of their self-destruction, something that finally resulted in a continuous increase of the quantities of the produced particles and antiparticles.

And whereas the above processes went on in a Cosmos of charged particles, in a state of ataxia and instability and in an environment of charges and sub-masses –notions that we shall analytically discuss in our work "the creation of matter and antimatter"–, when an electron approached a proton the following might occur, exactly as it happened with the pointons and anti-pointons:

a) The electron might fall on the proton due to the attraction exerted between them, resulting in the self-destruction of the two particles.
b) The second thing that might happen was that the electron might pass very far from the proton and nothing would occur, except that the electron would draw away normally and continue its course among the other particles and
c) The electron might pass at such a distance from the proton that nothing of the above would happen, but it would be trapped by the proton and start a rotating orbit around it. So, in this way the first atom of hydrogen was formed, as shown in Figure 17.

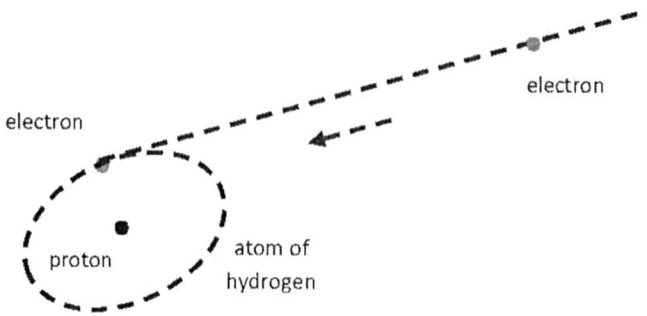

Figure 17. The creation of the atom of Hydrogen.

In the third case, the electron would be transformed from an enemy of the proton to its guardian and it would protect it from the other particles and mostly from the other electrons, which would no more be able to approach the proton as they were repulsed by the electron.

With the formation of those complexes, which essentially were the atoms of Hydrogen, a new stage started to appear gradually, that of the creation of masses and the stabilization of the Universe.

The most reasonable evolution after the creation of the first atoms of Hydrogen was the creation of other atoms of Hydrogen in exactly the same way, in a number that would reach not only the number of the atoms of Hydrogen in our Universe, but in the whole Cosmos.

Now, along with the formation of the atoms of Hydrogen, approximately equal quantities of atoms of Antihydrogen were formed too. This means that when an antielectron entered a rotating orbit around an antiproton an atom of Antihydrogen was formed and, gradually, along with the atoms of Hydrogen, a respective quantity of Antihydrogen atoms was formed.

At the same time as the formation of the atoms of Hydrogen and Antihydrogen, which in fact represented the first particles of matter and antimatter, the formation of the atoms of Helium started, in exactly the same way. The atoms of Helium were formed when free electrons entered rotating orbits around the respective nuclei of Helium that had already been formed during the fourth stage of the creation. The ratio of the

Helium atoms were created to the Hydrogen atoms was about one atom of Helium for three atoms of Hydrogen. In parallel to the formation of the atoms of Helium, respective quantities of antiatoms of this element were also created in exactly the same way. At this point it has to be noted that the other nuclei and the other atoms that are present in the Universe and the Antiuniverse were created at their largest part much later, within the Stars

The atoms of Hydrogen and Helium formed had two properties different than those of the charged subatomic particles that had been formed, i.e.:

The first different property of the atoms of Hydrogen and Helium compared to the charged subatomic particles was that they had neither charge nor polarity and behaved like neutral particles among the other particles. However, a very small and imperceptible, almost zero, attractive interaction developed between them. This imperceptible interaction was the gravitation as it accepted today by physics. The same imperceptible attractive interaction developed also between the atoms of Antihydrogen and Antihelium. This force became repulsive between the respective atoms and antiatoms.

The second different property of the atoms compared to the other particles was that at the very small atomic distances the imperceptible attraction of the gravitation became zero –instead of increasing– and then became repulsion. The theory of the chain reaction named this property *"dynamic equilibrium of the mass"*. This second property, combined with the first one, consisted of the elements that contributed in the creation of the masses, the material bodies and then the creation of the whole Universe.

These two properties were so imperceptible that, initially, and even possibly until now, they were not considered as worthy elements to become the subject of a specific study or investigation and it should be noted that the second property, of the dynamic equilibrium of the mass, remains unknown as an atomic property until today and is mentioned for the first time in the theory of the chain reaction.

The creation of the atoms and antiatoms of Hydrogen and Helium completes the fifth stage of the creation and, essentially, completes the theory of the chain reaction about the microcosm. The next three stages we describe in the next chapter concern the megacosm and the cosmos of the infinity and are part of the theory of the creation of matter and antimatter; it might be said that they consist of a supplement of the theory of the chain reaction, about the megacosm and the cosmos of the infinity.

So, upon the end of the fifth stage of the creation, we have a Cosmos consisting of a huge cloud formed at a certain spot in the infinite, absolutely empty space, which was a mixture of pointons, quarks, electrons, protons and neutrons, nuclei of Helium, atoms of Hydrogen and Helium, along with their respective quantities of antiparticles. This cloud will be called in this work, from now on *"cosmogonic gas"*.

We can characterize the five stages described as stages of "ataxia and instability"; the first one as the stage of "charges"; the second, third, fourth and fifth as the stages of "submasses", names that will be rationally explained in the theory of the "creation of matter and antimatter", described in the third book of the trilogy.

Now, the formation of the atoms and antiatoms of Hydrogen and Helium puts an end to the stages of "ataxia and instability" and the stages of "stabilization" or stages of the "masses" start, as then the neutral atoms and antiatoms formed, began to join and separate in stable groups of atoms and stable groups of antiatoms. These groups, as we shall see, formed the matter, the antimatter, the "Universe", the "Universes", the "Antiuniverses" and the whole "Cosmos".

In general, the rest of the evolution, developed in the same way as the respective periods in the theory of the "big bang" with certain differentiations concerning mostly the formation and separation of matter and antimatter and the case of the creation of not only our Universe, as it accepts the theory of the "big bang", but and other Universes and Antiuniverses as well, which as a whole formed the Cosmos as described in the next chapter.

In fact, of course, the stages of ataxia and instability, as well as all the stages of the creation continue until now and will continue in the future at more intense and ever increasing rates. However, we use the expression that the stages of ataxia and instability ended in contradistinction to the beginning of the development of the new stages, that is, the stages of the stabilization of the Universe, as we shall analytically discuss in the next chapter.

Someone might surely wonder how it is possible for a "Universe" and then a "Cosmos" to be created inside a gas where pointons, quarks, electrons, protons, atoms of Hydrogen and Helium coexist with all their antiparticles. However, afterwards, the events did not evolve exactly like this, neither did free charged and neutral particles coexist or matter and antimatter coexist. We shall describe the exact way in which the events developed in the next chapter, that about the creation and the separation of matter and antimatter and the creation of the Universe, the other Universes and Antiuniverses and the Cosmos.

COMMENTS ABOUT THE ESTABLISHMENT OF THE ROTATING ORBITS OF THE POINTONS ANTIPOINTONS AND QUARKS INSIDE THE NUCLEUS: THE STRUCTURE OF THE HYDROGEN ATOM AND THE SUGGESTION ABOUT THE DEFINITE STRUCTURE OF THE ATOMS

Until now, physics has not been able yet to describe the way in which the elementary subnuclear particles move inside the nucleus. As we have already discussed, concerning the motion of pointons and quarks inside the nucleus, the "theory of the chain reaction" supposes that these particles, during their interactions for the production of other particles, move on rotative orbits around their opposite particles, similarly to the motion of the electrons around to the opposite charged nucleus.

At this point a question arises: is the assumption that the particles in the nucleus move on rotating orbits indeed correct and, if so, how is it possible that all these movements are performed harmoniously in such a limited space as the nucleus?

The answer to this question is based on the rationale that, as we accepted that the electrons perform rotating orbits around the nucleus, there is no reason to not accept that the pointons and the quarks rotate too around their particles with an opposite charge –and/or vice versa–, forming thus the protons, and the neutrons.

Now, in what concerns the issue of the space needed for the above movements, if we compare the subnuclear dimensions calculated in the third book of the "trilogy of the creation" to the dimensions of the atom, we shall note that there is total proportion in size, speed and space that permits the harmonious realization of all the above rotating motions of particles inside the nucleus too. Such an assumption, compared to the subnuclear dimensions calculated, may be considered absolutely normal.

On the contrary, we can consider fairly arbitrary the assumption that, it is possible to have space inside the atom for the rotating motion of the electrons, without having the respective space for the same motion of the particles within the nucleus too, given that we consider that inside the nucleus we start with particles that have even zero size.

At the same time, taking into account that the speeds of these particles approach the speeds of the electromagnetic radiation, we may say that, with the establishment of the rotating orbits of the subnuclear particles, the "theory of the chain reaction" settles a problem that preoccupied scientists for many years. This happens because it is impossible to think that there could be particles moving in linear orbits in the limited space of the nucleus or in the space of a neutron or a quark.

Based on the above rationale, the "theory of the chain reaction" accepts that the particles move in rotating orbits not only inside the atom but inside the nucleus too, at all stages, immediately after the formation of the elementary particles, point charges, pointons and antipointons. In this very point, I would like to note too, that the frequency of the above rotations is so big that it gives the impression that these particles do not behave as spots points, but form the impression that there is a cloud around the center, of their motion.

The remarkable fact in this case is that the acceptance of the rotating orbits for the particles inside the nucleus, conforms and combines very efficiently with the requirements of the behavior of the particles during the initial stages of the creation, so that their motion obey both, the natural laws and be completely conformed to the subsequent developments. At the same time, however, the acceptance of the rotating orbits of the particles inside the nucleus enables us to provide answers to many unanswered questions, concerning the subatomic particles, such as: How particles with mass and dimensions, were created, from particles without mass or dimensions? Or how is the coexistence and coherence of the sub-nuclear particles within the nucleus achieved?, etc.

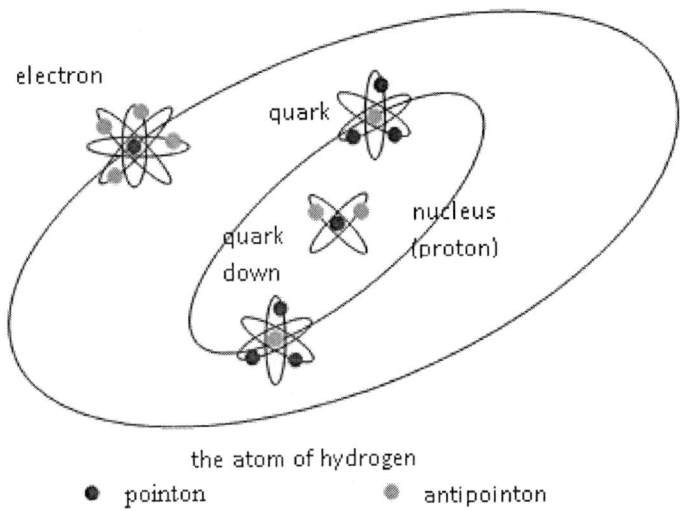

Figure 18. The suggested indicative the structure of the atom of hydrogen, consisting of combinations of elementary particles pointons and antipointons.

After the adoption of the rotating orbits for the particles of the nucleus, an indicative structure can be suggested for the atom of Hydrogen similar to that shown in Figure 18. But we must note at this point of our study, that the structure of the quarks, protons, neutrons, electrons and atoms of Hydrogen, suggested in Figures 15, 16 and 18, respectively, must be considered just indicative. We temporarily establish those indicative structures for those particles in order to describe the creation of the other particles from the pointons and antipointons, the production of the strong nuclear force and gravitation, the function of the dynamic equilibrium of the nucleus, the dynamic equilibrium of the mass, etc.

However, the configuration and the establishment of the definitive structure of the particles, as these will result after the acceptance of the rotative orbits inside the nucleus, will have to be done with great caution and after the absolute completion of the study about the unification of these particles, so that their structure will be totally identical with that defined by the currently available theoretical and experimental data, and at the same time they should be fully conformed to the real and natural behavior of these particles in the microcosm.

For the "theory of the chain reaction", the details of the structure of the particles at this state of study, are not as important as the understanding of certain basic principles of the theory. For instance, the exact structure of the proton, the neutron and the electron is not as important as the suggestion that these particles are *"divisible particles"*, given that their charge is multiple the charge of the smallest subdivision of charge that has been traced. The suggestion that the creation of matter begins from particles, *"just charges"*, without mass and without dimensions is also important.

Of great importance is also the suggestion about the "electromagnetic interaction" between pointons and antipointons and that this is the *"only"* interaction in the Cosmos and the suggestion that the rotating orbits of the particles inside the nucleus start immediately after the creation of the elementary particles pointons and antipointons. Finally the importance of the very satisfying way in which the above concepts combine both among themselves and with the principles described in chapter ten must be emphasized, so that they will provide in the end a uniform theory about the unification of the fundamental forces and the physical theories.

Certainly, in the "theory of the chain reaction", the above concepts do not concern only the subnuclear particles, the nuclei of Helium and the atoms of Hydrogen and Helium that we have seen until now, but extend also to all the existing particles, as well to all atoms.

Thus, the picture of the atom according to the "theory of the chain reaction" has the general and probably even the definitive form depicted in

Figure 19. The figure also shows the series of subatomic and subnuclear particles created during the first five stages of the creation; from the creation of the elementary particles, to the creation of the atoms; namely:

In the first line of the explanations of the symbolism of the particles, is the pointon and antipointon, the first stage of the creation.

In the second line, the quark up, the quark down and the electron, the second stage of the creation.

In the third line, the proton and the neutron, the third stage of the creation.

At the center of the atom, the nucleus, the fourth stage of the creation.

And final Figure 19, which presents the atom, the fifth stage, which is and the last stage of the ataxia and instability.

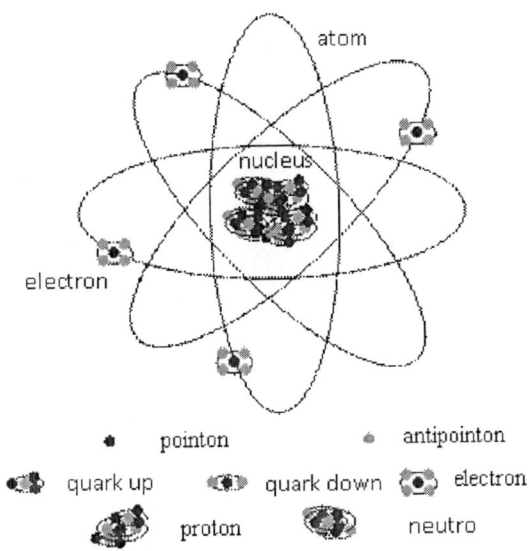

Figure 19. The structure of the atom according to the theory of the chain reaction.

Chapter 7

THE CREATION OF OUR UNIVERSE THE OTHER UNIVERSES THE ANTIUNIVERSES AND THE COSMOS: THE SEPARATION OF MATTER AND ANTIMATTER. THE COSMOS IN ITS ACTUAL FORM ACCORDING TO THE THEORY OF THE CHAIN REACTION

Newton's theory in itself is not enough to describe the Universe, because in the big cosmic distances the relativistic phenomena are differentiated and become of great significance. Therefore, in our investigation, we should indeed use Einstein's "general theory of relativity".

However, as it is found, Newton's theory is surprisingly successful in the study of several of the questions about the dynamics of the Universe and, therefore, in this reading we shall be able to manage things fairly well without using the "general theory of relativity".

Certainly, there are many questions about the geometry of the Universe, its size and shape, which "Newton's theory" cannot answer.

From Hans C. Ohanian's *Physics II* optional reading

At this point, the author of the above reading let us comprehend that besides "Newton's theory" that describes fairly well the phenomena of the Universe, there is also the "general theory of relativity", which describes in more detail the points that Newton's theory fails to describe or leaves obscure. However, he does not provide specific information about these points or how the theory of relativity describes them.

I personally believe that a clear and fairly daring opinion is that the general theory of relativity can give us no answers at all, about the questions of the creation as it has never been completed as a theory and, in general, failed to express a clear, simple and reasonable conclusion about the structure and the creation of the Universe. I also believe that Newton's theory is absolutely valid, but only for the megacosm, that is, the cosmos of the closed distances, i.e., for dimensions bigger than the atomic ones, but smaller that the astral, heliacal or galactic ones.

The view I formulate above is fully explained in the "trilogy of the creation", where, in the theory of the unification of the fundamental forces and the physical theories, we replace or, better, we complete Newton's gravity and the law of the Universal attraction with the Universal gravity that applies to the whole Universe; where the validity of the Universal gravity starts from the elementary particle charges, pointons and antipointons, and extends to the infinity.

THE "COSMOGONIC GAS"

In the previous chapter, we left a Cosmos consisting of the time, an absolutely void space of infinite size and a huge quantity of cosmogonic gas, in a certain region of this space. As described, the components of the cosmogonic gas were a mixture of pointons, quarks, electrons, protons, neutrons, nuclei, atoms of Hydrogen and Helium and all the antiparticles and antiatoms, of these particles and atoms.

Certainly, there were also several other particles, nuclei and atoms in the cosmogonic gas, but in much lower proportions, and played in general a secondary role in the creation of the Cosmos. The most important property of the cosmogonic gas was that, as a total energy inside the Cosmos, it had zero total energy and that any partial quantity of the gas, as a subtotal, had zero energy too.

In general, we can divide the components of the cosmogonic gas in two big classes. These classes are the charged subatomic particles and antiparticles and the neutral atoms and antiatoms. The interactions between the charged particles and antiparticles that formed the cosmogonic gas were attractive when the charged particles and antiparticles were heteronymous and repulsive when they were homonymous. These interactions increased when the charged particles approached and became infinite when they came into contact, resulting in their self-destruction.

The interactions of the neutral atoms and antiatoms differed at certain points from the interactions or the charged particles and antiparticles. One of the points where the interactions of atoms and antiatoms differed was that these were very weak interactions, about 10^{-36} times weaker than the interactions of the charged particles and antiparticles, so weak that they were perceptible only with large concentrations of atoms or antiatoms.

Another property in which neutral atoms and antiatoms differed from the subatomic charged particles and antiparticles interactions was that these weak interactions between atoms and antiatoms were always attractive. On the contrary, the interactions between atoms and antiatoms were interactions of the same power but repulsive.

These interactions became more powerful when the atoms or the antiatoms approached, exactly as it happened with the charged particles. However, contrary to the charged particles, when the atoms or the antiatoms approached more than a certain distance, these interactions, instead of becoming infinite, became zero.

The above different properties in the interactions of atoms and antiatoms are those that created gravity and masses. As we have already described, the "theory of the chain reaction" named these interactions gravitational inter-actions. These correspond to the gravitational forces, as these are currently accepted by physics.

The interactions between the charged particles and neutral atoms, follows the properties of the interactions of neutral atoms. All the above interactions between the components of the cosmogonic gas are summarized in Table 3, on the following page.

Table 3. The interactions between the elements of the cosmogonic gas

	positively charged particles	negatively charged particles	positively charged antiparticles	negatively charged antiparticles	neutral atoms, neutrons	neutral antiatoms antineutrons
positively charged particles	repulsion electro/tic interaction	attraction electro/tic interaction	repulsion electro/tic interaction	attraction electro/tic interaction	attraction gravitational interaction	repulsion gravitational interaction
negatively charged particles		repulsion electro/tic interaction	attraction electro/tic interaction	repulsion electro/tic interaction	attraction gravitational interaction	repulsion gravitational interaction
positively charged antiparticles			repulsion electro/tic interaction	attraction electro/tic interaction	repulsion gravitational interaction	attraction gravitational interaction
negatively charged antiparticles				repulsion electro/tic interaction	repulsion gravitational interaction	attraction gravitational interaction
neutral atoms, neutrons					attraction gravitational interaction	repulsion gravitational interaction
neutral antiatoms antineutrons						attraction gravitational interaction

Notes:
a) The interactions between charged particles and antiparticles become infinite when particles and antiparticles come in contact. On the contrary, the interactions between atoms and antiatoms become abruptly zero and then are reversed when atoms and antiatoms approach more than a certain distance.
b) In the case of the neutron particles, the interactions are different than the interactions of the other charged subatomic particles and antiparticles. The properties of the differentiated interactions, of neutron particles, generally follow the properties of the interactions of neutral, atoms and antiatoms. Analytical details about the interactions of the atoms, the neutrons and their antiparticles are given in the third book of the "trilogy of the creation", where the theory of the creation of matter and antimatter is discussed.

The Creation and the Separation of Matter and Antimatter the "Sixth Stage of the Creation"

The properties of the elements of the cosmogonic gas, as described in the previous section, rendered atoms and antiatoms able to form stable groups of atoms and stable groups of antiatoms within the cosmogenic gas.

Indeed, the atoms that behaved as neutral elements did not react with the charged particles of the gas and thus they were not mutually destroyed. Their only interaction in the cosmogonic gas was the gravitational interaction, which produced an attraction between the atoms or between the antiatoms and repulsion between the atoms and the antiatoms. The result of this behavior was the formation of the stable groups of atoms and the stable groups of antiatoms mentioned above.

Then, these groups developed in individual and stable clouds of matter and stable clouds of antimatter, consisting basically of atoms and antiatoms of Hydrogen and Helium, respectively. However, due to the fact that the gravitational forces between the atoms of Hydrogen and the atoms of Helium were different, the clouds of matter formed consisted of clouds with separated quantities of Hydrogen and separated quantities of Helium.

So, at the end of the sixth stage of the creation, we have a Cosmos consisting of the vast void space, the cosmogonic gas in which clouds of Hydrogen of matter, clouds of Hydrogen of antimatter, clouds of Helium of matter and clouds of Helium of antimatter, were formed.

The Accumulation of Matter and Antimatter and the Formation of Quasars – Protogalaxies – the "Seventh Stage of the Creation"

The sixth stage of the creation, as described in the previous section, lasted several million years. However, the exact duration of this stage is of no interest, as the "theory of the chain reaction" has totally disconnected the stages of the creation from their duration. In any case, in fact, the theory of the chain reaction accepts that there are no time boundaries in the duration of each stage, given that even the initial stages continue until now at the boundaries of the Universe, the Universes and the Antiuniverses.

However, what we can say is that the stages of the creation according to the "theory of the chain reaction" evolved very quickly, almost in no time. Based on rough calculations, I can note that the period needed from the creation of the first pointon until the creation of the first antipointons was no more than 10^{-36} seconds. The first quarks and the first electrons had already been formed 10^{-34} seconds later and the first protons, neutrons and the initial nuclei 10^{-32} seconds later. After 10^{-30} seconds, the first atoms of Hydrogen and Helium had already been formed too. However, the "theory of the chain reaction", as we have already explained, does not need the durations of the stages; in any case, it is too early for study this issue.

But, besides the disconnection of the stages from their duration, the reader may have noted that, in the theory of the chain reaction, although we have already reached the description of the seventh stage of the creation, we have not mentioned the word "temperature" yet. Is this a mistake or an omission in the theory or something else? The answer to this question is very simple. There is neither a mistake nor an omission in the theory; nor does anything else happen. The "theory of the chain reaction" is simply not based, at least for the initial stages of the creation, on the differentiation of the temperatures, as it happens with the theories that have been suggested until now. We describe and explain the reasons that make the "theory of the chain reaction" not to need the differentiation of temperatures for the description and the evolution of the stages of the creation of the Universe in the second and third books of the "trilogy of the creation".

But let us put a full stop in this parenthesis about the duration and the temperature in the various stages of the creation and proceed to see what happened after the formation of the clouds of matter and antimatter that had been formed into the cosmogonic gas.

The dynamic properties of the clouds were the following: attraction between the clouds of matter or between the clouds of antimatter and the repulsion between the clouds of matter and antimatter. These properties, in time –in a period estimated now in several million years–, provided the possibility for the formation of large concentrations of Hydrogen and Helium that created the Quasars, –very brilliant celestial bodies, with intense radiation, noted today at the boundaries of the Universe–, the Protogalaxies, –i.e., Galaxies in a primitive form, which formed our initial Universe– and the Stars, of the Universes.

At the same time with the concentrations of Hydrogen and Helium, large quantities of Antihydrogen and Antihelium were also formed, which created the Quasars, the Protogalaxies and the Stars, of the Antiuniverses, respectively.

THE FORMATION OF THE SOLAR SYSTEMS AND THE GALAXIES, THE "EIGHTH STAGE OF THE CREATION"

The eighth stage is, we might say, the stage during which our Earth, our Solar system, the other Solar systems, our Galaxy, the other Galaxies, the Universe, the other Universes, the Antiuniverses and the Cosmos as we know it today including all those mentioned in this paragraph were formed.

The eighth stage has started several million years ago and is the stage of the current cosmological evolution. The most characteristic feature of this stage is the element of *"Life"*, identified to the creation of entities that are aware of the Universe. We are not going to examine what is life and how it was created more than this brief reference, because it is a vast issue beyond the scope of this work, which studies exclusively the creation of the Universe from the material point of view.

The only thing I can say about life is that its creation and its evolution, although they took place in an infinitesimal part of the Cosmos, in our Earth, it still is the second most remarkable event after the beginning of the creation of the Cosmos, the second event after the creation of the initial particles-charges, pointons and antipointons.

So, Copernicus and Galileo may have proven that the Earth has nothing to do with the center of the Universe from the material and dynamic aspect, but from the aspect of *"Life"* it may even consists the center of the Cosmos.

I shall put an end to the description of the stages of the creation with the feeling and my belief that the "theory of the chain reaction" provides several clearer answers to the questions about the creation of the Universe, the Universes, the Antiuniverses and the Cosmos.

Of course, the theory adds, in turn, several new questions, such as, "What happens with the other Universes and Antiuniverses?", "Will we be ever able to explore another Universe or another Antiuniverse?", "Shall we ever reach the regions of the 'cosmogonic gas'?", etc., but I shall not proceed with my rest thoughts as I fear that then I shall never complete this work.

Table 4, on page 119, provides a brief picture of the evolution of the stages of the creation according to the "theory of the chain reaction" with the comment that, in the "theory of the chain reaction" these stages, as we have already described, are not characterized by periods of time or the prevailing conditions of pressure or temperature, but are exclusively characterized by the object of their processes.

The Cosmos in Its Actual Form According to the Theory of the Chain Reaction

After what we discussed in the previous sections of this chapter, we can conclude that the Cosmos in its current form and according to the "theory of the chain reaction" should be depicted as follows:

Initially, it consists of a vast theoretically infinite, absolutely void, bound-less space that ends nowhere. In a region of this space, the cosmogonic gas is produced in enormous quantities, which occupy a huge space, but although infinitesimal compared to the existing absolutely void and unused space.

Inside the cosmogonic gas the Universe is found, exactly as we are aware of it today, with dispersed but finite boundaries, enclosed in the regions with the cosmogonic gas. In cosmogonic gas many other Universes are also found, similar or different than ours. There is also a respective multitude of Anti-universes. The Universes and the Antiuniverses are fueled with new matter and antimatter, which is continuously produced in the cosmogonic gas. In parallel to the evolution of the existing Universes and Antiuniverses, the creation of new Universes and new Antiuniverses begins. During the initial stages of their creation, the new Universes and the new Antiuniverses created, move at very low speeds, which constantly increase due to the repulsion between matter and antimatter, until their motion reaches the speed levels noted today.

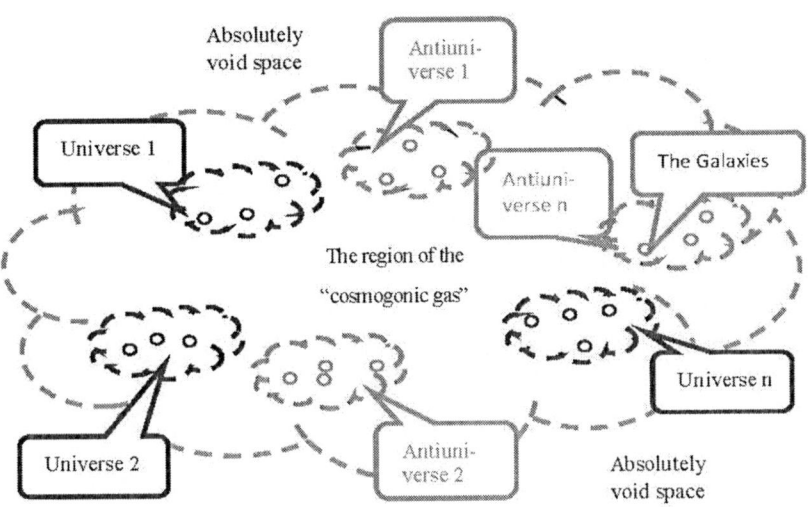

Figure 20. The Cosmos in its final form, according to the theory of the chain reaction.

So, the possible picture of the current Cosmos may be depicted in the form of the Figure 20, where the absolutely void spaces of the Cosmos, which have no boundary and reach infinity, the region of the cosmogonic gas, the Universes, the Antiuniverses, the Galaxies in the form of small spots and the voids among the Universes, the Antiuniverses and the Galaxies are shown.

Of course, the picture of the Cosmos as depicted in Figure 20 is very simplified. In fact and according to the "theory of the chain reaction", the Cosmos is much more complicated than that. For instance, we might say that there are Universes and Antiuniverses totally different in size. One, two or more Antiuniverses can correspond to each Universe and/or vice versa. The shapes should be much more irregular than shown in Figure 20. At the same time, the boundaries of the Universes, the Antiuniverses and the cosmogonic gas should be much more confused and obscure. Also, the positions, the motions and the distances of the Universes and the Antiuniverses should be very different and much more complicated, etc.

For the creation of the Cosmos, two main factors have contributed and continue to contribute up to day. These factors are:

- *The chain reaction, which creates the elementary particles pointons and antipointons and*
- *The electromagnetic interaction between the pointons and antipointons, which generates the electromagnetic force.*

These two factors create the rest of the other particles and antiparticles the other fundamental forces and everything else that contributes to the creation and the functioning of the Cosmos.

Anyway, recapitulating, we may say that today we live in a Cosmos that begins from the elementary particles, pointons and antipointons and extends to enormous distances that reach infinity. In this infinite Cosmos, with the infinite dimensions, the Universe, the Universes and the Antiuniverses consists huge subsets compared to the dimensions people are used to. However, these subsets are infinitesimal compared to the size of the whole Cosmos.

The above theoretical description is an attempt to explain the creation of the Cosmos, based on the existing data, on logic, clear, simple and correct thoughts, with no irrational assumptions or arbitrary axioms. That is the reason I believe, it describes and the real way of the creation of the Cosmos; but let the rest,... be judged by the reader,... after studying the theory.

...and concluding the only I can say...

"That is the Cosmos and so it been Created. !!!"

Table 4. The stages of the creation according to the "theory of the chain reaction"

1st Stage	2nd Stage	3rd Stage	4th Stage	5th Stage	6th Stage	7th Stage	8th Stage
The creation of the pointal charges pointons and antipointons	The creation of the quarks and antiquarks, the electrons and the antielectrons	The creation of nucleons (protons and neutron)	The creation of the nuclei of Helium	The creation of the atoms and antiatoms of Hydrogen and Helium	The creation and the separation of matter and antimatter	The creation of the Quasars, the Stars and the Prtogalaxies	The creation of the Solar systems and the Cosmos in its current form
The stage of the charges	The stages of the submasses ||		The stages of the masses			
The stages of ataxia and instability (coexistence of matter and antimatter)				The stages of stabilization (separation of matter and antimatter)			

Note: According to the theory of the chain reaction, the stages of the creation of the Cosmos continue until today at the boundaries of the Universes and the Anti-universes at an increasing rate. This is why the "theory of the chain reaction" separates the stages according to the object of the process of each stage and not based on their duration or the temperature, as in the theory of the "big bang".

Chapter 8

A BRIEF, IMAGINARY, THEORETICAL JOURNEY, TO AN ANTIUNIVERSE

One can imagine two kinds of Cosmos, the first like the one we live in, where positive electricity is related to the atomic nucleus, around which the electrons spin, and a second one in which the nuclei are negatively charged and the positrons spin around them.

Peter Debye, Nobel Prize in Chemistry, 1936

The Cosmos of antimatter should have the same characteristic features as our Cosmos. Undoubtedly, the matter of this Cosmos will be stable just as the matter of our Cosmos.

From an article by Dimitrios Kostakis,
Professor of Astronomy at the University of Athens, 1963

The "theory of the chain reaction" incorporates the above presumptions about antimatter in a reality.

THE PREPARATION AND THE BASIC RULES FOR OUR JOURNEY TO THE ANTIUNIVERSE

Humanity is in a several hundreds of years, posterior of this century and ISCR –a hypothetical society of cosmic research, the International Society of Cosmic Research– prepares its first manned journey to an Antiuniverse. This journey is one stage of a project for the exploration of the Cosmos, which is now in progress.

The project includes the completion of the exploration of several solar systems in certain galaxies of our Universe and a preliminary simple visit to an Antiuniverse by an unmanned and then a manned spaceship. This mission of the manned spaceship will complete the ISCR project.

The largest part of the project has already been completed with total success. The mission of the unmanned spaceship to an Antiuniverese has preceded and now the last part of the project remains to be executed, that is the mission of a manned spaceship to this Antiuniverse.

I believe Jules Verne would start his narration in such a way if he were alive today and tried to write a novel about an imaginary journey "From the Earth to an Antiuniverse" instead of the novel "From the Earth to the Moon". We shall describe this journey to the Antiuniverse in a similar way, but with the following differences. Jules Verne was very well aware of the fact that his writings were creations of his imagination, regardless if many of those fiction works were then verified as true events. On the contrary, however, we added in the title of our journey the word "theoretical", using this definition in order to emphasize that our descriptions, although imaginary, are based on very sound theoretical scientific assumptions and therefore there is a very big chance that they end in real events.

However, despite the fact that our descriptions have sound theoretical possibilities to be at the same time real as well, it is very difficult if not impossible to verify them with the current scientific data. The too long distances to the regions where the described events develop, consist the reason that renders their verification difficult, as opposed to the very low speeds that we can actually use in order to visit these regions. In fact, we need speeds exponentially higher than the speed of light!!!

The above uncertainty about the effectiveness and the success of the efforts paid by science in finding experimental data about what happens beyond the boundaries of our Universe and up to the other Universes and Anti-universes is the reason why I chose to describe this journey in this way.

However, as the scope of this chapter is not the writing of a novel with nice expressions and fictional descriptions but the theoretical description of the above according to the "theory of the chain reaction", I shall try to make it as short as possible completing our entire imaginary but theoretically documented journey in just a few pages. I personally believe that this journey is absolutely real but will "never" take place.

Two young cosmonauts are going to participate in this journey. The first will be responsible for the operation and the navigation of the spaceship and the second will proceed to observations during the journey in the Universe and then in the Antiuniverse. Of course, a role interchange between the two cosmonauts is foreseen in case of emergency.

The construction of the spaceship that will perform this journey has already been completed and its launching has been planned for the next days, as soon as the weather conditions in the region of the launching base will allow it. ISCR, as contractor, named the spaceship "Pioneer" thus wishing to remind us that "Pioneer" will be the first manned spaceship to visit an Anti-universe.

"Pioneer's" construction is totally different than the construction of the spaceships constructed and used by people in the early third millennium, as many of the features of those spaceships have been now significantly improved.

For instance, the fuel used now occupies minimum space, the navigation and observation instruments have been perfected and the construction materials, as well as the comfort are by now appropriate for very long journeys realized at speeds multiple of that of the speed of light!!!.

We shall use two new hypothetical devices for the navigation and the observations inside the Antiuniverse, which at this stage of the exploration are limited in simple visual observations. The first is a device similar to spectacles, constructed especially for the observations inside the Antiuniverse. ISCR named this device "antimatter spectacles".

The second device, which supplements the imaginary equipment of our spaceship, is the matter and antimatter particles detector. The matter and anti-matter particles detector is a high-tech device that allows us to trace and identify all the particles of matter and antimatter that exist in the Cosmos.

A Description of the Currently Prevailing Ideas in the Sciences of the Microcosm and the Cosmos of Infinity: We Are Suppose to Be in a Posterior Century Nobody Knows How many Centuries Later

However, during the waiting for the amelioration of the weather conditions in the region, so that they will allow the launching of the "Pioneer", we shall proceed to a brief flashback to the advances made in the past centuries in science, the cosmos of the infinity and the microcosmos. First, the cosmological theories that prevailed in the beginning of the third millennium were not able to answer several basic questions and they are gradually abandoned. Today very few scientists refer to those theories. In any case, those theories described the creation of just one Universe, without foreseeing the creation and the existence of many Universes and Antiuniverses. So, in this case, our current journey to the Antiuniverse would make no sense at all.

New theories have replaced the former ones, describing in a more understandable and convincing way the cosmos of the infinity and the microcosm. The concept of the "Universe", as it was meant in the end of the 20th century has been for a long time replaced by the concept of the "Cosmos", a concept by which we finally refer to the totality of the Universes, the Antiuniverses, the regions of the cosmogonic gas –that is beyond the boundaries of the Universes and the Antiuniverses– and all the void and absolutely void spaces, that extend to infinity.

At the same time, the theories of relativity were not able either to convince us that the speed of light is the highest speed in the Universe and that it remains the same for all observers and for all the inertia systems – something that consisted the base of the theories– or to explain the causes that might connect the speed of light to the energy of the relativistic equation $E = mc^2$ and the mass or to describe gravity in a convincing way. So these theories were gradually abandoned as cosmological theories and remained as simple interesting physical theories. Certainly, if the theories of relativity were valid and the highest speed in the Universe remained at the levels of the speed of light, we would not be able to realize this journey to the Antiuniverse, as if we used a spaceship moving at speeds close to that of light we would need at least 2.10^{10} years for this journey, something inconceivable based on what data is known today.

On the opposite side, for the exploration of the microcosm, science has explored the depths of the nucleus and considers it a fact that the particles "pointons" and "antipointon" are elementary particles and constitute the smallest subdivisions of matter and antimatter. At the same time, the theory of the Quanta has been considerably modified, especially in the field of the particles of interactions, which have been found to be simple interactions indeed and that all the interactions are caused by only one interaction, the electromagnetic interaction taking place among the elementary particles pointons and antipointons. This means that all the interactions, i.e., the electromagnetic, the gravitation and the weak nuclear interaction, the interactions of light, heat, etc., included, are all grades of the electromagnetic interaction. Today the "theory of the chain reaction" that was formulated in the beginning of the 21st century, which starts from the elementary particles and leads us to infinity, is the one considered valid as a theory for the creation not only of the Universe, but of the entire Cosmos. The theory has been verified experimentally too in its largest part.

Certainly, several modifications and improvements have been made and incorporated in the original formulation of the theory, improving it considerably. Also, from the 21st century until now, many other cosmological theories have been suggested, but none of them has been able to justify its cause and displace the theory of the chain reaction. In any case, the theory of the chain reaction describes the creation of the Universe and the entire Cosmos so convincingly, simply, precisely and reliably that there is not even the thought for a serious modification or replacement of the theory.

Concluding this little intervention, I wanted to make it clear that in my descriptions the use of speeds higher than the speed of light is a completely fantastic and no scientific idea, since the limit of the speed of light is a world speed limit. Thus, while the descriptions of our trip to an Antiunicerse have very large theoretical bases and can, somehow prove to be true descriptions, speeds will in no case overcome the speeds of light. This is why in this chapter references to speeds greater than the speed of light are always followed by three exclamation points!!!

THE LAUNCHING AND OUR TRIP, TO THE BOUNDARIES OF THE UNIVERSE

However, the weather conditions improved considerably and thus the procedure for the launching of the "Pioneer" is situated in the past. At this very moment "Pioneer" has already left behind the Earth, the Moon, the planets of our solar system and travels towards the vast space at a speed that has exceeded that of light by far!!!. "Pioneer", in order to be able to respond to the time schedule ISCR has planned for its journey, according to which it must reach the boundaries of our Universe one month after launching, has to develop a speed of at least 2.10^{11} light years per hour!!!. At this point, the reader can very well understand the difficulties this journey would encounter if the spaceship traveled at a speed of one light year per hour. Essentially, it would be as if we wished to realize this trip by

A Brief, Imaginary, Theoretical Journey, to an Antiuniverse 127

an almost stationary spaceship! However, the technology of the last centuries allowed us to construct spaceships moving at very high speeds, such as those mentioned in the above paragraph, and thus we had the opportunity to plan this journey. Pioneer has already reached the speed of 2.10^{11} light years per hour!!! and after having passed from several Galaxies heads for the extremities of our Universe. During the journey in the Universe, besides testing the "antimatter spectacles" and the "matter and antimatter particles detector", the cosmonauts had no other specific significant mission. Several supplementary observations constituted a simple control of the results of some previous missions. The only strange thing during our journey in the Universe was that when observing it with the antimatter spectacles our observations were different than those made with common spectacles. The colors and the brightness of the stars seemed rather different. Even the black color of the sky was quite different too. We have already reached the thirtieth day of our journey and according to the distances we have traveled we should approach the boundaries of the Universe. We fastened our seatbelts for the case of any extreme weather phenomena that might develop upon our entry in the region of the cosmogonic gas and continued our course.

THE PASSING THROUGH THE REGIONS OF THE COSMOGONIC GAS

We have reached at the fiftieth day of our journey and we travel inside the regions of the cosmogonic gas. These regions received us rather indifferently in general; we might describe their reception as if absolutely nothing happened or as if they did not realize our presence at all. Anyway, we almost too, did not realize when we left the Universe and entered the regions of the cosmogonic gas. The boundaries between the Universe and the cosmogonic gas were so confused and extended in so immense distances that only the relative observations we made with the matter and antimatter particles detector confirmed that further from a certain point we traveled inside the cosmogonic gas.

At first, the particles detector traced the presence of quarks, antiquarks, pointons and antipointons, particles that as we are aware of exist in very large quantities only in the cosmogonic gas. It also started to trace smaller quantities of electrons, neutrons, protons, hydrogen atoms, helium atoms, along with approximately equal quantities of their antiparticles.

As we advanced in our way inside the region of the cosmogonic gas, large quantities of atomic particles started to be detected and we noted concentrations of matter of Hydrogen and Helium gases, along with respective and almost divided between them concentrations of antimatter of the above gases.

At the same time we encountered also very large –endless– regions of voids. We had been familiarized with the differences between voids and concentrations of matter and we were able to distinguish them even visually, without the use of instruments, i.e., when we passed from voids of concentrations of gases of matter and antimatter, the scenery changed significantly. It is worth mentioning that when we passed from regions of antimatter we saw the opposite image from the one in the regions of matter. Thus, when observing the regions of antimatter with the antimatter spectacles, its image was the same with that of the regions of matter without the spectacles. All those were very interesting, but strange as well!

At very far distances we observed scattered and very sparse, several luminous bodies that the people at the ISCR had explained were primitive creations of Primary Quasars, Quasars, Red giants, Stars, Solar systems and Galaxies.

By intuition, we had also learned to distinguish whether these were bodies formed by concentrations of matter or antimatter. In any case, in general, excluding the presence of multiple subatomic particles of matter and anti-matter and the relatively small distances between the concentrations of gases of matter and antimatter, the image was not very different from that seen during our journey inside the Universe. Concerning the term "Primary Quasars", a term ISCR used for the first time, they had explained to us that it refers to quasars in which matter and antimatter have not yet been totally separated.

It seems, however, that the size of the regions of cosmogonic gas was much bigger than those estimated by Cosmology. We have already reached the hundredth day of our journey –at speeds much higher than those reached at the boundaries of the Universe– and based on the indications we have, we believe that we may still be at the center of a region of the gas.

During the previous days we encountered regions with large voids, where we noted scant particles and interactions, as well as regions of atomic particles of matter or antimatter; regions with quite strong concentrations of protons, electrons and quarks and large aggregations of Hydrogen and Antihydrogen, Helium and Antihelium, similar to huge pools –with sizes multiple that of the size of our Universe– that fueled Solar systems, Galaxies, even Universes and Antiuniverses. When we were inside regions of matter we sensed a very friendly environment compared to that of antimatter, which seemed annoyed by our presence and tried to throw us out of its region.

Several days passed and the above scenery –large and smaller concentrations of gases of matter and antimatter, atoms of Hydrogen and Helium, subatomic particles, etc., alternated continuously. Sometimes we detected also several different particles or atoms, but in very small quantities. ISCR had warned us about such phenomena and so differentiations like that did not impress us specifically.

OUR ARRIVAL AND OUR ITINERARY
INSIDE THE ANTIUNIVERSE

The fuel and food supplies in the spaceship had been calculated for a journey of about eighteen months and we had already reached the seventh month of our journey. We were approaching the safety limit of our supplies and our fuels, considering that we needed about the same time for our return journey. We had already started to prepare a return plan, taking into account a safety period of time for eventual unforeseen incidents. What we wondered most was whether we were going to encounter a larger region of antimatter, larger than the ones we had encountered inside the regions of the cosmogonic gas, which might probably lead us to the Antiuniverse.

Indeed, at a certain moment we noted a decrease in the detections of subatomic particles, which meant that we would have again a certain alternation of situations. Day by day the subatomic particles became increasingly less until they disappeared in the end. The pointons and antipointons were the first to disappeared and then the quarks and all the other particles.

Luminous bodies of antimatter started to appear far in the horizon, just as we were able to discern them, and gradually, as we approached, they were more easily visible and in larger quantities. The dispersion of those bodies was generally similar to the dispersion of the Stars of our Universe. By simple instinct we realized we had reached our destination.

Indeed, when we put the antimatter spectacles on in order to better watch the environment surrounding us we saw an incredible view. It was just if we were in our own Universe! Nothing was different.

Through the antimatter spectacles, the Antiuniverse looked exactly like the Universe. It was so identical that at first we thought we had made a mistake in our course and had returned to the Universe. So, the Universe and the Antiuniverse are completely identical and did not differ but in the way in which we looked at them.

However, many of the "Pioneer's" instruments gave us false indications; therefore, in order not to be influenced by eventual false observation, we had switched them out of operation. The instruments that functioned normally were the antimatter spectacles and the elementary particles detector, which confirmed that we were in an opposite Universe, i.e., an Antiuniverse, indeed.

Everything we watched in the Universe with naked eyes we watched also now in the Antiuniverse with the antimatter spectacles. The particles detector detected particles of antimatter instead of particles of matter.

A Brief, Imaginary, Theoretical Journey, to an Antiuniverse

We continued our way in the Antiuniverse and using the instruments we had in active mode we observed many stars, large concentrations of antimatter containing quasars, stars, solar systems, galaxies, etc. Our way was extremely interesting but at a certain point we had to start the journey back. The fuels of the spaceship had already reached marginal quantities. So, because of the time pressure too, we decided to approach the closest solar system and visit the closest star of the system we would encounter in order to study the behavior of antimatter on our spaceship; then we would start our journey back.

With this goal in mind, we reached the first planet of the Antiuniverse. We named this planet, which was the first star of the Antiuniverse encountered by humans, "Metic". Metic was similar to the planets of our solar system from various aspects. In general, we might say that it was an amalgam of the planets of our solar system, Pluto, Neptune and Uranus, but it also had many other characteristic features totally unknown to us.

Upon our arrival to Metic the most basic part of our mission started; that is, the study of the behavior of antimatter on matter. Indeed, when we tried to put Pioneer on a rotating orbit around Metic, we noted that we had to proceed to the exactly opposite manipulations compared to those we did to maintain a satellite on a rotating orbit around the Earth or a spaceship on a rotating orbit around a certain planet or star in the Universe. This happened because Metic exerted a repulsive force instead of an attracting one. In other words, Metic exerted repulsion (negative attraction) on the Pioneer. This repulsion was, in general, the "antigravity"!

After quite a long time we learned that the announcement of the above data to the ISCR resulted in intense enthusiasm and great satisfaction. Indeed, almost without realizing it, we had conducted the biggest experiment in our century, the "antigravity experiment". According to the instructions ISCR had given us, this experiment would mean the end of our trip. We made some rounds around Metic in order to admire the environment and greet in our own way our arrival in the Antiuniverse and prepared for the journey back to Earth.

THE RETURN FROM THE ANTIUNIVERSE

We drew our route towards our Universe and turned, leaving Metic behind; gradually, it disappeared in the horizon. We did not need any further effort in order to start our way back, as Metic's antigravity assisted to that, giving us the impression that the planet, although not hostile, was not very hospitable and it was like as if it wanted to drive us by any means, but absolutely politely, away from its surroundings. The return journey was in general the same with the going, except that the fuel consumption of our spaceship during the way back was much less than the original consumption. The repulsion of the Antiuniverse assisted that very much, as well as an almost imperceptible attraction by the Universe that probably tried in this way to bring us back as soon as possible. As we had not gone far into the Antiuniverse and we were near its boundaries, we did not need much time to reach the regions of the cosmogonic gas. Then, after crossing the regions of the gas we entered the Universe and headed towards our Galaxy and then toward our Solar system. We passed several Galaxies and when we entered our Solar system we headed towards the Earth. The landing, which is conducted today just as airplanes land in simple airports, –I remind the readers that we are supposed to be at the end of a posterior century–, took place in front of a large crowd that had been gathered at the landing base and wished to hear our experience from this journey. In parallel to Pioneer's landing, which meant also the end of this imaginary journey in the Antiuniverse –a journey that is never going to take place due to the immense distances in which the events described develop compared to the very low speeds of the spaceships in our possession–, we too landed in our current reality.

COMMENTS, REMARKS AND CONCLUSIONS FROM OUR JOURNEY TO THE ANTIUNIVERSE

I believe that all of us wonder, more or less, whether the Universe in which we live is the only one or if we live in a Cosmos with multiple Universes. I also believe that we have all wondered about antimatter and the existence of an Antiuniverse or even more Antiuniverses. If we proceed to a mental review of our knowledge about the above, we shall find that until reading the theory of the chain reaction, besides the words Universes, antimatter, Antiuniverse and Antiuniverses we did not know and had not heard anything more about those concepts. The most basic questions arising from the above concepts are: what is antimatter, how it is created and where might the other Universes and Antiuniverses are?

Many scientists tried to describe the concepts of antimatter, the Universes and the Antiuniverses in order to provide some answers to the above questions. However, these descriptions represent rather imaginary or even philosophical thinking, as no one among these scientists has incorporated his description in a scientific theory in order to justify his views based on theoretical data. Based on this rationale, I feel that the theory of the chain reaction is the first theory that explains and describes in a specific way the creation of matter and antimatter, from which the Universe and the other Universes and Antiuniverses were then formed.

However, according to the theory, the other Universes and Antiuniverses must be very distant from ours. These distances are in the range of very high and most probably exponential multiples of the size of our Universe and thus we should be talking about huge distances between our Universe and the other Universes and Antiuniverses, distances almost imperceptible compared to the current sizes of our Universe.

At the same time, many scientists and physicists as well have accepted as a doctrine the view that, in order to be established, a theory must be verified experimentally too. However, in the case of multiple Universes and Anti-universes an experimental verification might be impossible due to their inconceivable distances. So, the question rise whether we should accept or not the existence of Universes and Antiuniverses and the notion of the Cosmos –that is, the total of the Universes and the Antiuniverses–, without an experimental verification.

The above question is very basic and has to be answered with great caution and wariness, as, if we accept the existence of Universes and Antiuniverses as an answer, even without an experimental verification, we might establish something that does not really exist. On the other hand, if we exclude the existence of Universes and Antiuniverses due to the fact that it cannot be experimentally verified, we shall probably be excluding a really existing concept. In this case and considering the above it will be very difficult to verify the truth and we may never be able to do it.

So, for the above reasons, I suggest a median course of thinking that I consider the golden mean for the problem. This is the way of thinking our ancestors, the Ancient Greeks, adopted when they established the first scientific views about the Universe. This rationale is the following: A theory has to be based always on clear theoretical and experimental data. However, if we are not able to collect adequate experimental data, but the theory is based on sound theoretical bases, we have to accept it until the refutation of the bases that support it, even without any experimental verification.

Certainly, the physical theories about the megacosm are usually simple theories that are often verified by only one experiment. For example, I would mention the theories about heat transmission, the motion of a body in a gravitational field, the inclined plane, the flow of a fluid, etc. However, the cosmological theories about the cosmos of infinity and the theories about the microcosm are complicated theories and so one experiment or even a limited number of experiments might constitute the necessary only and not the absolutely sufficient data that would found these theories.

So, a cosmological theory, in order to be established, besides by experiments, must be also based on very sound, clear and understandable theoretical views. In this case, that means, when the experimental data are not sufficient, the conformity of the theory with the laws and the principles of nature must be taken very seriously into account and the theory should be established only when all the laws, principles and axioms connected to it and supporting it have been clarified.

In our days, the two previous paragraphs are overlooked resulting rather in the opposite; cosmological theories are established, based on deficient and disputable –probably even erroneous– experiments without any convincing theoretical support.

So, as I mention also above, I feel that the theory of the chain reaction is the first theory to be founded on very rational, documented and clear theoretical perceptions, whereas it is based on the fact that all natural laws are applied regularly from the very first moment of creation. At the same time, the theory describes the formation of the elementary particles and analyzes in detail the way in which matter and antimatter and then the multiple Universes, Antiuniverses and the Cosmos were formed.

Undoubtedly, the experimental verification of the multiple Universes and multiple Antiuniverses would be a very precocious attempt. The effort for a very precise description of antimatter and the Antiuniverses is also very precocious. This is the reason why the theory of the chain reaction chose this imaginary theoretical journey to an Antiuniverse for the theoretical description of what is taking place beyond the boundaries of our Universe.

Chapter 9

THE FUNDAMENTAL UNANSWERED QUESTIONS ABOUT THE CREATION: THE ANSWERS OF THE THEORY OF THE CHAIN REACTION

    I thought very hard before deciding in which way I should write this chapter. My personality leads me to write only the positive points of my views. And then I prefer to let the readers draw their own conclusions about what already exists.

    However, my faith in the "theory of the chain reaction" –and perhaps, as I must admit, a certain nervousness about the correctness of the already existing theories, which I believe are in a false course of research–, made me write this chapter.

    In general, I feel that lately many theories about the creation of the Cosmos and our Universe come to light, but they can hardly be acknowledged as such. Often, these theories express so exaggerated views that instead of "cosmological theories" they should rather be called "science fiction views".

    So I believe that this chapter creates the conditions that will help us clarify the limits of a "theory" from those of a "science fiction view", as we shall see in the exposition of this chapter that in order to answer several questions we do not need so much imagination but rather correct scientific thinking.

Thoughts of the author

SOME BASIC QUESTIONS

The various cosmological theories provide many data about the creation and the evolution of our Universe. However, together with the valuable data they provide, the theories leave also unanswered many questions that today expect an answer from science.

We have already mentioned several of those questions in previous chapters and especially in chapter five in the description of the section about the reasons that established the "big bang". In the present chapter, after a recapitulation of the basic questions about creation that remain still without a convincing answer, we shall analyze how the theory of the chain reaction provides an answer for them.

In general, the most basic of those questions that expect an answer are:

How were the vast quantities of energy present in the Universe created?

How were the particles quarks formed? Are quarks elementary particles?

Do the particles-carriers of interactions exist and if so how were they formed and how were they incorporated in matter?

How were the nuclei of the atoms formed from the quarks?

How were gravitation, the masses and the material bodies formed?

How did the natural laws apply in the beginning of the creation?

Does antimatter exist and if so where is it?

From the above questions, I shall analyze only the first one, "How were the vast quantities of energy present in the Universe created?". Justifying this choice to analyze only the first question, I might say that although the other questions are very basic and have not yet found any convincing answers in the various cosmological theories, for the theory of the chain reaction the answers to them are so simple and provided in such a clear and understandable manner in its respective chapters that I believe give the reader the impression that these questions do not exist in this theory.

For instance, the answer provided by the theory of the chain reaction to the question, "does antimatter exist and if so where is it?", we may say that the answer of the theory is clearly positive and states as one of its obvious preconditions, that at the same time with the production of matter, antimatter is produced too. From matter and antimatter the Universe was formed and then, the other Universes, the Antiuniverses and the whole Cosmos were formed, and develop.

There are also various groups of questions about the creation, which have already been answered with the respective theories that have been suggested. One such group consists of the questions about the reasons that established the theory of the "big bang". In this case we shall only study the answers provided by the theory of the chain reaction to these questions, letting the reader judge and compare them to the already existing ones.

Another group of questions added to the above are those produced by the two basic goals of physics, i.e., the exploration and unification of the fundamental forces and the physical theories and the exploration of the mechanism of the formation of matter and antimatter. We describe elements of these explorations in the next chapter.

How Were the Vast Quantities of Energy That Exist in the Universe Formed?

We are all aware of the fact that we live in a Universe where vast reserves of energy exist. However, the cosmological theories are not able to provide a clear answer about how this energy was created. This is the reason why the theories, based on mostly obscure explanations consider that this energy preexisted in the Universe. Certainly, such an answer, instead of simplifying our question "How were the vast quantities of energy present in the Universe created?" complicates it even further.

None of the existing cosmological theories refers to the notion of a Cosmos that was created from zero, i.e., was created without the consumption of energy; of a Cosmos in which many other Universes and

Antiuniverses were formed along with the Universe in order to counterbalance all this energy of the Universe.

Regarding the origin of the energy, are there in the Universe, the theory of the chain reaction, contrary to the other theories, starts from the precondition that originally there was absolutely nothing in the Cosmos except for the concepts "space" and "time". So, according to the theory of the chain reaction, everything started from zero and, therefore, the initial energy in the Cosmos was zero. Then, the energy required for the formation of the Universe, the other Universes, the Antiuniverses and the Cosmos was zero again. This means that no energy had to be consumed for the creation of everything we see around us. Besides, even if energy was necessary it would not be created, as there was none, as the theory of the chain reaction assumes that the creation of the Cosmos started from nothing.

Of course, nature functions always based on the obvious axiom that it is impossible for either energy or matter to be created from nothing. So, how is it possible to have a Universe with those huge quantities of energy and matter that were created without any energy consumption? Perhaps this question is the reason that makes the cosmological theories assumes the initial presence of energy. However, they are not able to explain its origin.

Although it is indeed impossible that energy was created from zero, we have already described in the theory of the chain reaction that under certain conditions of absolute void or inside the cosmogonic gas, which is produced without energy consumption, subsets can be created from zero, as energy entities, each of which has some energy but as a total have a total energy equal to zero. The entities that consist of subsets with energy but have a total energy of zero start from the pairs of the elementary particles, pointons and antipointons, pass in the cosmogonic gas where all its subsets have an energy sum equal to zero, reach the Universes and the Antiuniverses and end in the Cosmos where the total energy is again equal to zero. So, the Universes and the Anti-universes, as individual energy subsets possess huge quantities of energy, but as entities they have energy sums equal to zero.

So, in what concerns the energy in the Universe, the theory of the chain reaction conforms completely with the reasonable laws of nature and by assuming the totals with zero energy provides a very simple, rational and convincing answer about the origin of the energy present in the Universe.

THE QUESTIONS ABOUT THE REASONS THAT ESTABLISHED THE THEORY OF THE "BIG BANG" AND HOW THE THEORY OF THE CHAIN REACTION ANSWERS TO THESE QUESTIONS

In the fifth chapter where we described the theory of the "big bang" we saw that it was based on the observation of the drawing away of the galaxies in the Universe and the indications about conformity of the theoretical and experimental percentages of Hydrogen and Helium that constitute the Universe, as well as the presence of dispersed light radiation emitted by matter when the Universe was at the age of about seven hundred thousand (700,000) years.

Regarding the motion of the Galaxies in the Universe, the theory assumes that this motion is due to the expansion that resulted from the big explosion. Certainly, the theory cannot convincingly describe how and why the Universe started to expand in such a rapid rate that approaches the critical rate of expansion and how this rate is maintained unaltered almost until today, after so many billion years from the moment of the big explosion. At this point it must be noted that if the Universe started to expand at a slightly lower rate it would have already collapsed before reaching its current size.

The theory of the chain reaction answers the comment about the drawing away of the Galaxies as follows: in chapter seven we saw that the Antiuniverse is formed along with the Universe and that these two are repulsed. This repulsion produced at first a common rate of drawing away, which is however increasing due to the continuous repulsion, resulting in the levels we note today. Hubble was the one that noted this drawing away of the Galaxies and then many scientists considered that it was due to the expansion of the big bang.

Regarding the conformity of the theoretical and the experimental percentages of Hydrogen and Helium in the Universe we might presume that many theories exist mentioning the same rate of production of the above elements, which means the above conformity, has no relation only to the theory of the big bang, but in many other theories. So we might say that in the theory of the chain reaction too, the rate of production of Hydrogen and Helium is such that conforms to the above percentages.

Now, regarding the question why we detect the same dispersed radiation in any direction in the Universe, as mentioned above, the theory of the chain reaction considers that the explanation of this phenomenon is drawn automatically and is very simple: according to the theory of the chain reaction, as the Universe and the Antiuniverse are inside the cosmogonic gas. We are currently measuring the radiation emitted by the particles of the cosmogonic gas, which many scientists consider as the radiation emitted by matter when the Universe was at the age of seven hundred thousand years.

During the recent years astronomers study a new fact for which the existing theories are asked to provide an explanation. This fact is that the Galaxies in the Universe do not draw away just at a steady speed, but this speed is accelerated. No clear and satisfactory explanation has been given yet for this fact, as it is totally opposed to the basic principles of most cosmological theories. Some scientists have tried to explain this phenomenon by introducing two new theoretical concepts, the concept of dark matter and the concept of dark energy. Nowadays enormous experimental effort is being made to identify these concepts, but so far there is no a positive effect.

In the theory of the chain reaction, the case of the acceleration of the drawing away motion of the galaxies is explained as a simple normal effect of the repulsion between matter and antimatter that results then in the dynamics of the repulsion between the Universes and the Antiuniverses. It is obvious that this dynamics produced the current speeds of the galaxies that according to the laws of motion of the bodies result in an accelerated motion.

Chapter 10

THE THEORY OF THE CHAIN REACTION THE TRILOGY OF THE CREATION AND THE UNIFICATION OF THE FUNDAMENTAL FORCES AND PHYSICAL THEORIES[3]

It would be very hard to formulate at once a complete and uniform theory about anything present in the Universe. Instead, science advance step by step, with the discovery of partial theories, each describing a limited number of phenomena and either refers to the effects of the other phenomena arithmetically, or overlooking them.

However, we hope that we shall discover someday a complete, consistent and uniform theory that will include all the individual theories and will not contain any arbitrarily selected parameters that will posterior modulate the consistency between the theory and the data obtained by observation.

---

[3] This chapter was written for the readers who want to know some more about the unification of the fundamental forces and physical theories. In any case, it is an introduction to the issue of the unifications.

The research about such a theory is called "unification of the physical theories". Einstein dedicated most of his late years in an effort to discover such a theory, but unsuccessfully.

From Stephen Hawking's
A Brief History of Time

Reading the above paragraphs someone draws the conclusion that today, according to Stephen Hawking, there has been no worthy connection of the various partial physical theories in a unified whole that would include them all and explain everything that takes place in the Universe. Furthermore to Stephen Hawking's view as expressed above, I believe that it is not necessary to explain all the individual phenomena that take place in the Universe at once in only one clear theory in order to "unify the physical theories". But taking into account the comparison I make below that we can compare the total of the physical theories to a tree with multiple branches corresponding to the various physical theories, with the root and the trunk to present the basic theories, we may achieve this unification by discovering the theories of the root and the trunk, that supports the various other physical theories.

So, the basic theories that will describe the root and the trunk of the tree will constitute the essential theories of the unification. Then, each partial theory will be incorporated in the tree, as a branch of the tree and thus, we shall progressively form the complete image of the tree. And even if we miss a branch, it will not be so disastrous; the important thing is that its root and trunk, to be correct.

Unfortunately however, the current situation gives me the impression that, by trying to explore some of the branches of the tree, which probably unjustifiably consider as basic branches, we have taken a wrong direction and as a result, we have missed the root and the trunk of the tree.

ABOUT UNIFICATIONS

One of the most serious issues, physicists are interested in today, is the unification of the fundamental forces and physical theories. Scientists believe that, besides the unification of the fundamental forces and physical theories, the answers to the issues of the unifications will help at the same time, in the clarification of several other unanswered questions, about the creation and the functioning of the Cosmos. In the following sections, after a briefly referring to some cases of unifications, we will describe briefly the contribution of the "chain reaction theory" and then, the contribution of the "trilogy of the creation", to these unifications.

So we might say that this chapter is a very brief introduction to the second book of the "trilogy of the creation", which together with the third book of the trilogy, as we have already mentioned in the Prologue, will describe the two theories: that of the unification of the fundamental forces and the physical theories and that of the creation of matter and antimatter, which completes the theory of the chain reaction.

Essentially, these two books will be two separate theories that will constitute the theoretical part of the theory of the chain reaction and will be addressed to those readers interested in reading the theoretical details of the unifications and the details of the creation of matter and antimatter from the elementary particles-charges, pointons and antipointons, which are particles without "mass" or "dimensions".

THE CONCEPT OF THE UNIFICATION OF THE FUNDAMENTAL FORCES

The various natural phenomena in the Universe are presented as independent phenomena. Sometimes, however, science is able to unify these natural phenomena in groups with common features. So, whereas some natural phenomena at first seem different, they are then proven to belong to a unified group; so "unification" means the thorough study of these natural phenomena in order to find their common features and the links that connect them and classify them in the same group.

It is assumed that the physical forces that act on the elementary particles from which matter consists and which on its turn forms the material bodies, the satellites, the planets, the Stars, the Galaxies and the Universe constitute one group of such phenomena. Among these forces, physics distinguished the following four forces as the most basic ones, as we have already described in the fourth chapter about, matter, elementary particles and fundamental forces. These forces are the gravitational force, the electromagnetic force, the weak and the strong nuclear forces, which due to their significance were called fundamental forces.

The effort for the study and classification of the fundamental forces is among the efforts made by scientists to explain and classify the various natural phenomena in groups of unified sets. So, the "unification of the fundamental forces" is the effort to explore, study and classify the above four fundamental forces or even all the forces in nature in a unified group. It is certainly obvious that the unification of the fundamental forces will automatically result in the classification and unification of all the natural phenomena related to these forces, something that will then help us very much in our effort to unify all the physical theories too.

THE UNIFICATION OF THE PHYSICAL THEORIES

Everything that occurs in nature has or must have a reasonable explanation. Based on this rationale, we can define a "physical theory" as the study, description, analysis and explanation of a natural phenomenon.

However, the explanation of a natural phenomenon is often very simple, as for example the explanation about the phenomenon of the formation of waves in the sea; sometimes it is more complicated, as it happens about the reason of Earth's movement around the Sun; and other times it is even more complicated, as in what concerns the question how is it possible that protons and electrons coexist harmoniously inside the atoms, but we know why they coexist. However, in some cases, the explanation of a natural phenomenon is so complicated that we have not been able to find its explanation yet. Such a case is the explanation about the cause that makes the material bodies be attracted between them.

It is believed that the Universe originated from a starting point in the infinite time of the past. So the various events and the developments succeeded one another until we reached the current result. However, the sequential manner of evolution leads us also to the conclusion that there is not only succession of the events, but interdependence as well. This means that each event is related to another somewhat similar or previous one, as for example during a tempest the lightning is directly related to the thunder, etc.

Scientists try to study this sequence and correlation of the natural events, which, although it should be studied without interruptions, in what concerns our relevant knowledge and study, is often interrupted. This whole procedure was named "unification of the physical theories".

In this way, we might compare the total of physical theories to a tree with multiple branches, which correspond to the various physical theories.

Unfortunately, however, the root of the tree, many spots on its trunk and several of its offshoots were lost during the depth of the infinite years of the past. These are the spots we are currently trying to study in order to yield a convincing explanation for the various events and the whole image of the tree.

However, during our study, we should not identify the notion of the unification of the physical theories with that of the unification of the fundamental forces. These notions are independent and completely different notions; but they are often confused due to the fact that the forces are involved in many, but not all, physical phenomena. For instance, we might mention that the forces are involved in the formation of the atom, but they are not involved in the phenomenon of the eclipse of the Sun or the Moon.

However, as the contribution of the forces is of basic significance in the most natural phenomena, it is believed –and that is probably a great truth– that by the unification of the forces, physics will have advanced very much in the effort for the unification of the physical theories too. This is also the reason why the above issues are often studied interconnected.

THE BASIC UNIFICATIONS

Until Newton's era, Earth's force of gravity and the celestial force of gravity were considered two completely different forces. Newton, however, by formulating the law of the Universal attraction managed to unify those two forces and prove that in fact they are just one and the same force. This unified force was named "Newton's gravity force".

One century after Newton, Faraday and Ampere, with similar efforts, achieved the unification of the magnetic and electrical force, which until then were considered two different forces, in just one force named "electromagnetic force". In the previous —twentieth— century the unification of the electromagnetic force with the weak —nuclear— force was studied with fairly encouraging results for the moment. Many prominent scientists worked in this field, among them, the names of P.A.M. Dirac, Pet. Higgs, Tom Kibble, Sh. Ab. Salam and St. Weinberg, are distinguished. And it is believed that in the next years the connection between these two forces will be completely clarified.

Then Einstein, with the general theory of relativity, introducing the concept of space-time provided a new perspective in the research concerning the gravitational force and its unification with the other fundamental forces. However, in the general theory of relativity and the introduction of the concept of space-time, studied only the result of the phenomenon of gravity, —as it happens also in the universal theory of gravitation—, probably more precisely and thoroughly than in law of the universal attraction, whereas the cause that results it, is not studied yet.

This is why Einstein tried to supplement the above gap and complete his study about the cause of the gravitational force with a new theory on which he worked for several years. However, the results of his research concerning the unification of the gravitational force and the other fundamental forces were negative. The history of these unifications is depicted in Table 5, where the continuous lines refer to the already made unifications, the soft continuous lines, to the unifications are not completed and the dotted lines refer to the unifications that are disputed. The dates on Table 5 result from a simple review of the history of the fundamental unifications.

A very basic observation on Table 5 is that the unifications of the fundamental forces that were absolutely accepted and fully understandable are those in which the individual forces were completely identified, as for example the Celestial gravity and Earth's gravity and the electrical and the magnetic force. The unification of these forces is so complete that in the end, after their unification, they were given the same name.

Table 5: Time diagram of the basic unifications

Force / Date	Celestial gravity	Earth gravity	Magnetic force	Electrical force	Weak force	Strong force
18th Century	Newton					
19th Century			Faraday Ampè			
20th Century	Einstein			Glawhow, Salam, Weinberg, Higg etc		
21st Century			?			

So, I believe that the unification of the fundamental forces, as well as the So, I believe that the unification of the fundamental forces, as well as the unification of all the forces, will be successfully completed by the theory that will prove that all the forces of nature obey in one and only one principle and that these forces may differentiate at certain points but are essentially just one force. Following this rationale, I believe that the theory of the chain reaction, which from one aspect is based on only one interaction, the electromagnetic interaction, forms a very sound foundation for the above unification.

Now, regarding the unification of the physical theories, I believe that the theory of the chain reaction, after its completion, with the two theories, –the unification of the fundamental forces and the physical theories and the theory of the creation of matter and antimatter–, will contribute decisively to the discovery of the root and the trunk of the tree that depicts the various physical theories as described in the section about the unification of the physical theories. And then, it is suffice, that adding to the root and the trunk, those branches that we have studied or will study, will always have the image of the tree alive.

THE THEORY OF THE CHAIN REACTION AND THE UNIFICATION OF THE FUNDAMENTAL FORCES

So, regarding the organization and the functioning of the Universe, physics assumes that the four fundamental forces, –gravity, electromagnetic force, weak nuclear and strong nuclear forces– contribute to that, as we have already described.

These forces, acting like connective material and interact on the elementary particles of matter, which are the foundation stones, forming the masses, the material bodies, the Stars, the Solar systems, the Galaxies, the Universe, the Universes, the Antiuniverses and the Cosmos.

As we have already described in the previous sections, about the nature, the causes and the origin of the fundamental forces, the electromagnetic force, partially the weak and very slightly the strong nuclear forces have only been studied yet. Science has not been able yet to determine the cause of the gravitational force and define its relation to the other fundamental forces. However, the fact that gravitation is the most basic and the most commonest force in the megacosm and in the cosmos of the infinity, it creates several obscurities, which I feel will hinder many of physics current achievements until the discovery of the cause of gravitation. And I personally believe that not only do we face several obscurities in the whole structure of the fundamental forces, but we may be on the wrong track for the resolution of those obscurities.

But let us leave for a while the fundamental forces as science currently assumes them and proceed to a review in order to remember how the theory of the chain reaction accepts those forces. In the theory of the chain reaction, we accepted at first the force produced by the electromagnetic interaction between the elementary particles, point charges, pointons and antipointons. This force is exactly the same with the *electromagnetic force* as accepted it today by physics. Then, with the acceptance, by the theory of the chain reaction, of the fact that the rotating orbits of the subatomic particles continue inside the nucleus too and that they originate from the elementary particles, pointons and antipointons, formulated the quarks and the electron based only on the electromagnetic force. Through the quarks, –up and down–, reached to the formation of the protons and neutrons without needing any other force except the force produced by the electromagnetic interaction.

Now, the nuclei of the atoms, which consist of protons and neutrons, were formed with the support of a downgraded electromagnetic interaction, the *"strong nuclear force"*, as it is currently accepted by physics. Then, no other additional force was necessary for the formation of the atoms from nuclei and electrons, except the force of the electromagnetic interaction between the positively charged nuclei and the negatively charged electrons.

However, upon the formation of the atoms, a new downgraded force was produced, different in power but of the same origin with the electromagnetic interaction. This force was the *"gravitation"*, which, from the atoms formed the molecules, the masses, the material bodies, the Stars, the Solar systems, the Galaxies, the Universe, the Universes, the Antiuniverses and the Cosmos.

In essentially, the strong nuclear and the gravitational forces are two different grades of the electromagnetic interaction. These grades are produced when subatomic particles enter in rotating orbits around the particles of opposite charge; in particular when the quarks up enter in rotating orbits around the quarks down –or vice versa– in order to form the protons and the neutrons, producing the strong nuclear force or when the electrons rotating around the nuclei to form the atoms, producing the gravitation.

With the establishment of the electromagnetic force, the strong nuclear force and the gravitation, Table 5 on page 149 takes the final and definitive form of Table 6 on the previous page. Ending the description of this section I would like to point out that the theory of the chain reaction considers the strong nuclear force and the gravitation as the primary grades of the electromagnetic interaction and all the other interactions –light, heat, magnetism, weak nuclear force, etc.–, as the secondary grades of the electromagnetic interaction.

We describe the details of the mechanisms that produced the strong nuclear force and gravitation, the nuclei of the atoms, the atoms and the antiatoms, the matter and the antimatter analytically in the second and third books of the "trilogy of the creation".

Table 6. Time diagram of the unification of all the interactions according to the theory of the chain reaction

Interaction \ Date	Electromagnetic interaction "Electromagnetic force"			
	Primary grades			Secondary grades
	Gravity Celestial and earthly	Strong nuclear force		Weak nuclear force, light, heat, magnetism, etc.
18th Century	Newton			
19th Century			Faraday Ampere	
20th Century				Glashow, Salam Weinberg, Higg, etc.

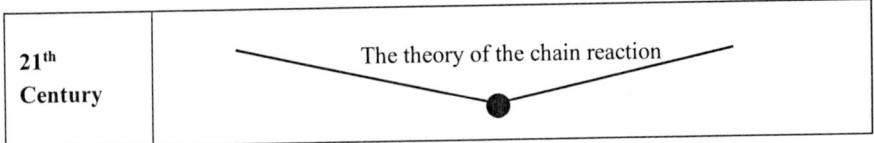

THE CONTRIBUTION OF THE TRILOGY OF THE CREATION IN THE UNIFICATION OF THE PHYSICAL THEORIES

In 1928 Max Born during a speech at the University of Gottingen stated that physics will come to an end in a period of six months. Max Born's optimism was based on the discovery of the equation of the behavior of electrons, made at that time by Dirac. So he hoped that a similar equation would be discovered very soon about the behavior of protons too, which were the only known particles then, resulting in the end of theoretical physics.

However, things became more difficult when it was discovered that protons were not the only particles in the nucleus, but that there were also the neutrons and the nuclear forces that connected the protons with the neutrons. But the difficulties did not end with the neutrons and the nuclear forces discovered, as they became tougher when it was found that the protons and the neutrons were, in their turn, divisible particles and that they consisted of triads of other particles, which were named quarks.

So, scientists started a new struggle to explore what is going on with the new particles and the new forces discovered and then to study the behavior of the new particles and the new forces based on the already existing data.

Einstein dedicated the last years of his life to an effort to discover a unified theory about the formation of the Universe, but with no success. The conditions at that time might not have been very favorable for such a research, as very little was known about the subatomic particles and the nuclear forces, which had just been discovered. We should only point out that when Einstein started his research neutrons and quarks had not yet been discovered.

New theories were formulated for the study of the new data; from those, "quantum mechanics" and the "theories of relativity" were distinguished. The theory of the big bang was based on these theories. However, scientists have not been able yet to unify these theories in a unified entity or connect them to the already existing basic theories about the functioning of the Universe, as for instance the universal theory of gravitation or the electromagnetic theory, etc. in order to form then a combination that would provide in turn a unified theory explaining in a convincing way what takes place in the Universe.

So I feel that the theory of the chain reaction will contribute decisively in the unification, as well as the sorting out of the existing physical theories and this without the need for any particular effort. This holds because the basic theories complement the theory of the chain reaction, i.e., the theory of the unification of the fundamental forces and the physical theories and the theory of the creation of matter and antimatter, originate from the very core of the theory of the chain reaction. So the notion of unification is automatic. In fact, we have only one unified theory that was simply divided in three in order to be better formulated and understood.

In this case, if we review the components that formed the theory of the chain reaction we shall find that it originated from the basic question, "what happens with the quark particles?" Are they indivisible? Or they are divisible particles? However, the characteristic charge of the quark up, which is equal to 2/3 of the charge of the proton, and that of the quark down, which is equal to 1/3 of the charge of the electron, lead us, as we have already discussed, to the conclusion that the quarks are, in their turn, divisible particles too. So the elementary particles "pointons" and "antipointons" were established, with a charge equal to 1/3 that of the proton and 1/3 that of the electron, respectively. As we have already described, the pointons and antipointons are elementary particles with neither mass nor size.

The selection of the pointons and antipointons as elementary particles proved to be a very successful one selection as then, with the completion and the acceptance of the electromagnetic interaction exerted between the above particles and assuming that the rotating orbits of the subatomic particles do not stop in the electrons but continue inside the nucleus too, – an absolutely natural assumption if not necessary too–, all the necessary and absolutely appropriate elements were formed for the formulation of the theory of the chain reaction.

After the selection of the pointons and antipointons and the establishment of the electromagnetic interaction, the next step in our research was to see how particles with mass and size could be formed from these particles that were simple charges, with neither mass nor size, and how the other interactions were produced from only one interaction, the electromagnetic interaction. This research led us to the formulation of the two adjoining theories, i.e., the theory of the unification of the fundamental forces and the physical theories and the theory of the creation of matter and antimatter, which with the theory of the chain reaction are linked in a unified entity that will constitute the "trilogy of the creation".

In this case, the reader can easily realize that the "trilogy of the creation" will not consist of three individual independent theories joined in one, but will essentially be only one theory, divided in three for typical reasons. So we see that there is absolutely no reason to think about a unification of the three theories composing the trilogy as they are interconnected and the unification is automatic.

At the same time, as we shall see in the analytical description of the unification of the physical theories, the trilogy meets all the requirements to incorporate all the other basic theories in a unified entity that will explain convincingly what takes place in the Universe. For example, I mention that the universal theory of gravitation is included as a whole in the theory of the chain reaction and is clarified for the case when the factor G for the universal gravitation is not absolutely constant but is minimally modified in very large distances and becomes zero or even negative in the very small atomic distances. Also, regarding the theory of electromagnetism, we may say that it is completely incorporated in the theory of the chain reaction as the theory starts from the elementary particles, pointons and antipointons. Pointons and antipointons are particles only electromagnetic charges.

So we can assume that probably the research for the unification and the explanation of what takes place in the Cosmos will not end with the "theory of the chain reaction" and the completion of the "trilogy of the creation", but I feel that, after the establishment of the trilogy, science will have put a very sound foundation for the unification of the physical theories and will have made a very big step to the right direction of its research and development.

EPILOGUE

I really felt great relief when I started writing the epilogue of this book. The reason I felt so relieved is that it meant I reached the end of an adventure that lasted many years, caused me a lot of wondering and deprived me of a lot of hours of leisure or another simpler activity.

I never expected that such a simple observation –that the quarks must be divisible particles–, or several simple thoughts –that the particles perform rotating orbits in the nucleus or that when a charged particle enters a rotating orbit around a particle with opposite charge the charges are not completely counterbalanced but a very small residue of effect remains that yield the strong nuclear force and gravitation– would lead me to such an adventure.

And I also never expected that my wondering would not end at the above thoughts but that, in order to be able to found them I would have to invent two new particles, the "pointon" and the "antipointon" from which I would have to remove every trace of mass and size and leave them only with their charges. And that I should then invent the "electromagnetic interaction" and "annul" all the particles of the interactions.

And certainly I also did not expect that in order to complete these thoughts I should undertake the writing of three books, –i.e., the "trilogy of the creation"–, formulating three complete theories, the "theory of the chain reaction", the theory of the "unification of the fundamental forces and the physical theories" and the theory of the "creation of matter and antimatter". And, at the same time, in order to explain the energy of the Universe, I had to create the Antiuniverse and the idea of multiple Universes and Antiuniverses, thus expanding the absolutely void space to infinity and creating the notion of the Cosmos.

And as if all the above were not enough, I then had to change the structure of the atoms, supplement Newton's laws and put again the mass that I had formerly removed to the other particles in order to form matter and, gradually, to form the Universe, the other Universes and Antiuniverses, which I placed swimming in the "cosmogonic gas", etc.

So how could I not feel such a relief, when realize that this adventure came to an end and that, before I start writing my new work, the unification of the fundamental forces and physical theories[4], I can take a short break. I wonder whether the result was worth that ordeal! But, because I have already finished writing this work, I shall not go on with my thoughts, but I shall let you judge, after thanking you for the interest, the understanding and your patience in reading of the book.

---

[4] The epilogue was written about ten years ago. Today, the second book of the trilogy of the creation: "The Real Grand Unification", in which I describe the unification of fundamental forces and physical theories, has already been released.

GLOSSARY AND A BRIEF EXPLANATION OF SEVERAL BASIC TERMS USED IN THIS BOOK

An italic font is used to explain the terms presented for the first time in this work.

Absolutely void space: *The space where neither matter nor interactions exist. This is the space outside the Galaxies, the Universes, the Antiuniverses, and the "cosmogonic gas". The absolutely void space expands beyond the "cosmogonic gas" and reaches to infinity.*

Antiatoms: Particles opposite to atoms.

Antiparticles: The particles of antimatter.

Antipointons: *Particles opposite to the pointon particles. They are dimensionless particles without mass and bear a charge equal to 1/3 of the charge of the electrons.*

Antiuniverse: At the same time with the formation of the Universe, the Antiuniverse is formed too, which is the opposite of the Universe.

Atom: The smallest mechanical subdivision of matter. Each atom consists of a positively charged nucleus, made of neutrons and protons, around which the negatively charged electrons spin. The number of the positively charged protons determines the chemical element to which the atom belongs, i.e., whether it is an atom of hydrogen, or an atom helium, or an atom of oxygen, or an atom of carbon etc.

Atomic distance: *A distance compared to that of the dimensions of the atom.*

Charged particle: A particle with positive or negative charge.

Copernican model: The Heliocentric model of our solar system. Nicolas Copernicus suggested the Heliocentric model of our solar system in the 16th century. The same model had been suggested in the ancient era, in the 3rd century B.C., by Aristarchus, but it was not accepted due to lack of arguments and means for its proof.

Cosmogonic gas: *According to the "theory of the chain reaction", it is a gas with zero energy found in huge quantities at the boundaries of the Universes and the Antiuniverses. It consists of a mix of pointons, quarks, electrons, protons, neutrons, nuclei, atoms of hydrogen and helium and the respective quantities of their antiparticles. Matter, antimatter and then the stars, the galaxies, the Universes, the Antiuniverses and all the Cosmos, are formed from this gas.*

Cosmos: *It is the total of the Universes, the Antiuniverses, the regions of the cosmogonic gas and the regions of the void and the absolutely void spaces. The Cosmos expands to infinity.*

Cosmos of the infinity: *The Cosmos of the very large distances. The Cosmos of infinity expands beyond our Galaxy and through the Universes, the Anti-universes and the absolutely void space reaches the extremes of infinity.*

Creation model: A set of theoretical rules, assumptions and parameters, which, adapted to the existing conditions is believed to be able to give us the real image of the creation and the functioning of the Cosmos.

Dynamic equilibrium of mass: *It's the property according to which, when the atoms get too close to each other, the "gravitational interactions", which according to the Law of the Universal Attraction should be infinite, are "on the contrary nullified".*

Dynamic equilibrium of the nucleus: It's the property according which, when the quarks get too close to each other, the "strong nuclear force", which according to Coulomb's law should be infinite, are "on the contrary nullified".

Electromagnetic force: The electromagnetic force is the one produced between two charged bodies.

Electron: This is one of the three basic –with the proton and the neutron– subatomic particles that form the atoms and then the matter. The electron has negative electrical charge.

Elementary particles: The subatomic particles constituting the last subdivisions of matter. The *"theory of the chain reaction"* accepts the entities *"pointon"* and *"antipointon"* as elementary particles.

Ether: A hypothetical substance, through which light is transmitted, as scientists used to believe. However the Michelson-Morley experiment proved that ether does not exist.

Fundamental forces: These are the basic forces participating in the basic functions of the Universe. These forces are four: the gravitational force, the electromagnetic force and the strong and the weak nuclear forces.

Galaxy: A group of stars and large concentrations of gases holding together under the effect of gravity. The distribution of the stars in the Galaxies is usually of elliptical or spiral shape. The number of stars in each Galaxy is usually in the range of one million to several billions.

Gravity –or Gravitation–: This is the attractive force exerted between two masses. Gravity as a natural phenomenon was first studied and described by Newton. Then, Einstein, with the theory of general relativity, tried to study the causes of gravity but without achieving any substantial results.

Helium: The second in quantity element, in gas form, in the Universe. Its weight corresponds to 23% of the Universe. In the inner part of the stars, where high temperatures and pressures develop, it fuses and turns into other, heavier elements.

Hydrogen: The element in the largest quantity, in gas form, in the Universe. Its weight corresponds to 85% of the Universe. In the inner part of the stars, where high temperatures and pressures develop, hydrogen fuses and turns into other, heavier elements.

Infinity: The concept of infinity determines what never ends.

Interaction: The remote effect between two entities through the absolute void, without interference of any substance, or other means, among these two entities.

Light year: The distance that light travels in a period of one year. This distance is approximately 9,460,000,000,000 kilometers.

Mass: The matter from which the various material bodies are formed.

Megacosm: The Cosmos of normal dimensions. That begins from the atom and ends at the extreme ends of our Galaxy.

Mental experiment: The theoretical model of an experiment that due to particular circumstances or to the means available cannot be practically performed.

Microcosm: The Cosmos inside the atom, within the nucleus of the atom and, according to the "theory of the chain reaction" even within the quarks.

Neutrons: Particles, which, with the protons, form the nuclei of the atoms. Neutrons are neutral particles without electrical charge.

Nova: Stars that in a very short period –due to nuclear processes in their inner part– increase in size and turn into red giants.

Point charge: Charge with point dimensions or rather a dimensionless charge. The elementary particles "pointons" and "antipointons" are point charges.

Pointons: They are the elementary particles from which, according to the "theory of the chain reaction", the creation of the Cosmos started. They are dimensionless particles without mass, with a charge equal to the 1/3 of the charge of the protons. They move with finite speed at the speed level of electromagnetic radiation.

Primary Quasar: Quasar in a very primitive form in which matter and antimatter have not been completely separated yet. We use this name for the first time in the "theory of the chain reaction". The "theory of the chain reaction" considers that Protogalaxies were formed from the primary quasars.

Protogalaxies: Galaxies in a very primitive form. Physics includes radiogalaxies too in this category. Currently, physics accepts that inside the protogalaxies even large quantities of antimatter exist. The "theory of the chain reaction" considers that protogalaxies are certainly aggregations of matter and antimatter that have not been separated yet.

Protons: Particles, which, with the neutrons, form the nuclei of the atoms. Protons are charged particles with positive electrical charge.

Quarks: Subnuclear particles from which the particles of the nucleus, protons and neutrons, are formed. Quarks have not been isolated as free particles yet.

Quasars: Very bright and very active concentrations of matter, of the size of the Galaxies, noted at the boundaries of the Universe. It is considered probable that the actual Galaxies and then the Universe originated from these concentrations.

Radioactivity: The radioactive fission of several atoms, as for instance of the atoms of uranium.

Relativity: In physics, the word relativity means the relation between events taking place in two or more different inert levels.

Sciences: These are the sciences studying mathematics, physics, cosmology, medicine, etc.

Group: A complete set of uniform elements or sets of uniform elements. For instance, the Universe is a group of galaxies and the voids between these galaxies.

Solar systems: Groups of limited number of stars in relatively short distances consisting distinct neighborhoods in the galaxies. Solar systems consist usually of a large self-luminous star –a star with its own light– like our Sun, around which other smaller stars, the planets, rotate. Often, other even smaller stars, the satellites of the planets, rotate around them.

Space-time: One of the most basic concepts of the theory of the general relativity. As the author of this book I ought to admit that even though I tried hard I never managed to accept or to understand the meaning of this concept.

Speed of light: The constant speed of light, which is represented by c and has the constant value of 299,792,458 m/sec all over the Universe. The theory of relativity considered that the speed of light is constant for all the observers regardless of their motion.

Star: A ball of mass consisting of atoms of hydrogen and helium, or other atoms, holding together by gravity. The stars tend to be grouped in formations that form the solar systems and then the galaxies.

Strong nuclear force: According to the current views of physics, this is the force that forms the subatomic particles, protons and neutrons, from quarks, and then holds these particles in the nucleus.

Subgroup: The division of a group.

Temperature: It shows how warm a body is.

Theory of the "big bang": A theory that describes the formation of the Universe. According to the theory of the big bang, the Universe, together with space and time, were formed by the explosion of a small ball with infinite energy, infinite temperature and infinite density. Although this theory explains many of the phenomena of the Universe, it fails to explain several basic points of the creation and, especially, where this small ball with the infinite energy, infinite temperature and infinite density that caused the explosion came from.

Theory of the chain reaction: This is a theory described in chapters six and seven of this book, which explains in a convincing, clear, complete and understanable way the creation of the Cosmos, from the elementary particles, to the infinity. It is based on the principle of the creation of the elementary point entities –with neither mass nor size– called "pointons" and "antipointons" and that only one interaction, the "electromagnetic interaction", takes place inside the Universe. The theory proves that all the other interactions are grades of the electromagnetic interaction. It also assumes that the elementary particles rotate inside the nucleus too. It is the first theory that explains how matter and antimatter were created, whereas it incorporates the notion of multiple Universes, Antiuniverses and the Cosmos. According to the "theory of the chain reaction", the absolutely void space extends to infinity.

Theory of the relativity: Einstein's theory, divided into two separate theories, the theory of the special relativity and the theory of the general relativity. The first is based on the assumption that the speed of light is always stable in all inert sets and for all observers, regardless of their mobility condition. This assumption yielded the well known equation of the equivalence of mass and energy, expressed by the well known formula "$E = mc^2$". The second theory tries to describe a relation between space, time and gravity, bur has never been completed. Regardless of the fact that these theories earned enormous publicity, they are now disputed by a large number of scientists. Regarding the first theory, the assumption that the speed of light is the same for all the observers is disputed; regarding the second theory, it has never been able to provide a rational explanation about the cause of gravity and connect this cause with the space and time, that, according to the theory created gravity.

Unification of the fundamental forces: This is the effort made in physics for the exploration, the study and the classification of the four fundamental forces of nature –the gravitational force, the electromagnetic force, the strong and the weak nuclear forces– as a unified entity.

Unification of the physical theories: This is the effort made in physics for the invention of a consistent theory that will connect all the existing partial theories, —each of which describes a limited number of physical natural phenomena—, in a unified entity.

Universe: All the Galaxies and all the void spaces. It is estimated that the Galaxies that form the Universe are several billion.

Void space: This is the space where no matter is present. Contrary to the absolutely void space, in the void space interactions exist.

Wavelength: The distance between two successive crests of a wave. The wavelength of a radiating body determines the spectrum of radiation based on which we can determine the speed and the material properties of the radiating body.

Weak nuclear force: The force yielding the radioactivity.

REFERENCES

To write this book, I used a very large bibliography that includes almost everything there is in this field of science. But mention only the basic bibliography I used:

Avison John: *"The World of Physics"*.
Alexopoulos K.: *"General Physics"*.
Chrysis G.: *"The Universe and the cosmological evolution"*.
Danezis-S. Theodosiou M.: *"The Universe I Loved"*.
Elbaz E.: *"The Quantum Theory of Particles"*.
Ford Kenneth: *"Contemporary Physics"*.
Grammatikakis G.: *"The Autobiography of Light"*.
Grammatikakis G.: *"Vereniki's Hair"*.
Hawking Stephen: *"A Brief History of Time"*.
Walter Tony Hey-Patrick: *"The Quantum Universe"*.
Ohanian Hans C.: *"Physics"*, Volumes I and II.
Salam Abdu: *"Unification of Fundamental Forces"*.
Singh Simon: *"Big bang"*.
Young Hugh D.: *"University Physics"*.

AUTHOR CONTACT INFORMATION

Vaggelis Talios
Mechanical and Electrical Engineer
National Technical University of Athens, Athens, Greece
Email: vtalios@gmail.com

INDEX

A

antimatter, 32, 44, 45, 61, 63, 66, 74, 79, 80, 90, 95, 96, 98, 99, 102, 103, 109, 112, 113, 114, 115, 117, 119, 121, 124, 125, 127, 128, 129, 130, 131, 133, 135, 138, 139, 142, 147, 151, 154, 156, 157, 160, 161, 162, 165, 167

antipointons, 54, 61, 62, 63, 69, 83, 84, 85, 86, 87, 88, 89, 90, 91, 93, 94, 95, 96, 97, 103, 104, 105, 106, 107, 110, 114, 116, 118, 125, 128, 130, 140, 147, 153, 156, 157, 159, 161, 163, 164, 167

Antiuniverses, 2, 4, 32, 34, 37, 53, 85, 102, 103, 109, 114, 115, 116, 117, 118, 124, 125, 129, 133, 135, 139, 140, 142, 152, 153, 160, 161, 162, 167

Aristotle, 9, 13, 23, 56, 57

ataxia, 99, 102, 107, 119

atmosphere, 24, 38, 47, 48, 49, 50, 51

atmospheric pressure, 47, 49

atomic distances, 101, 157, 162

atoms, 6, 30, 33, 34, 45, 53, 54, 56, 57, 58, 59, 60, 62, 66, 67, 68, 70, 75, 78, 91, 97, 98, 99, 100, 101, 102, 103, 104, 105, 106, 107, 110, 111, 112, 113, 114, 119, 128, 129, 138, 148, 149, 153, 154, 160, 161, 162, 163, 164, 165, 166

B

bible, 28

Big Bang, 8, 29, 31, 53, 71, 72, 73, 76, 77, 78, 79, 82, 102, 119, 138, 139, 141, 142, 155, 166, 169

black hole, 37, 43, 45

C

carbon, 44, 47, 77, 162

celestial bodies, 2, 13, 26, 43, 44, 51, 75, 115

chaos, 11, 12, 13

classes, 40, 41, 42, 50, 51, 111

classification, 41, 42, 148, 167

coherence, 45, 105

conformity, 134, 141, 142

consumption, 86, 132, 139

cosmogenic gas, 113

Cosmos, 1, 2, 3, 5, 6, 8, 11, 12, 13, 22, 30, 32, 33, 36, 37, 48, 49, 53, 54, 63, 64, 66, 73, 81, 83, 84, 85, 86, 87, 93, 99, 100,

102, 103, 106, 109, 110, 113, 115, 116, 117, 118,119, 121, 122, 124, 125, 132, 133, 134, 135, 137, 139, 140, 146, 152, 153, 157, 160, 162, 164, 166
cosmos of the infinite, xiii

D

direct observation, 10, 61
distribution, 52, 163

E

egg, 72, 73, 74, 78
electromagnetic, 34, 35, 64, 66, 67, 68, 69, 75, 84, 85, 87, 90, 91, 92, 93, 104, 106, 118, 125, 148, 150, 151, 152, 153, 154, 156, 157, 159, 163, 165, 167
electromagnetic charges, 157
electromagnetic interaction, 66, 68, 69, 85, 87, 91, 106, 118, 125, 151, 153, 154, 156, 157, 159, 167
electromagnetic waves, 84, 90, 91, 92
elementary particle, 1, 3, 8, 32, 48, 54, 55, 57, 58, 59, 60, 61, 66, 69, 74, 78, 84, 87, 88, 89, 91, 93, 96, 99, 104, 105, 106, 107, 110, 118, 125, 130, 135, 138, 140, 147, 152, 153, 156, 157, 163, 164, 167
energy, 3, 4, 34, 44, 68, 69, 72, 73, 74, 78, 83, 85, 86, 87, 110, 125, 138, 139, 140, 142, 160, 162, 166, 167
energy consumption, 140
environment, 23, 25, 92, 99, 129, 130, 131
equilibrium, 66, 68, 84, 98, 101, 106, 163
ether, 91, 92, 93, 163
evolution, 7, 8, 9, 11, 12, 25, 27, 30, 32, 33, 36, 56, 62, 66, 71, 74, 75, 78, 79, 85, 86, 98, 100, 102, 114, 115, 116, 117, 138, 148, 169

F

finite speed, 87, 88, 90, 165
fission, 33, 68, 165
fluid, 134
food, 12, 129
force, 3, 12, 28, 34, 35, 45, 48, 64, 66, 67, 68, 69, 70, 75, 78, 86, 88, 90, 91, 96, 98, 101, 106, 118, 131, 148, 149, 150, 151, 152, 153, 154, 159, 163, 166, 167, 168
formation, 3, 4, 8, 33, 45, 48, 61, 62, 63, 66, 67, 68, 69, 74, 75, 79, 84, 87, 90, 93, 95, 96, 97, 98, 99, 100, 102, 104, 113, 114, 115, 135, 139, 140, 148, 149, 153, 155, 161, 166
fundamental forces, 8, 55, 61, 63, 64, 66, 67, 69, 70, 78, 81, 106, 110, 118, 139, 145, 146, 147, 148, 149, 150, 151, 152, 156, 157, 159, 160, 163, 167, 169

G

galaxies, 2, 4, 29, 30, 31, 34, 37, 39, 44, 45, 52, 53, 64, 71, 75, 76, 77, 78, 80, 88, 115, 117, 122, 127, 128, 129, 130, 132, 141, 142, 147, 152, 153, 161, 162, 163, 165, 166, 168
Galileo, 27, 28, 49, 116
geometry, 17, 19, 109
gluons, 68, 74, 75
God, 11, 12, 13
grades, 125, 153, 154, 167
gravitation, 28, 34, 35, 48, 64, 66, 68, 75, 101, 106, 125, 138, 150, 152, 153, 154, 156, 157, 159, 163
gravitational constant, 64
gravitational effect, 45
gravitational field, 35, 134

Index

gravitational force, 34, 64, 65, 67, 69, 70, 91, 111, 113, 148, 150, 152, 153, 163, 167
gravity, 14, 24, 35, 45, 48, 49, 56, 64, 65, 67, 69, 78, 87, 110, 111, 125, 149, 150, 151, 152, 154, 163, 166, 167
Greece, 13, 22, 171
Greeks, xvi, 15, 16, 18, 19, 21, 22, 23, 134

H

harmony, 2, 13, 77
helium, 43, 46, 49, 50, 75, 76, 77, 79, 98, 99, 100, 101, 102, 103, 106, 110, 113, 114, 115, 119, 128, 129, 141, 142, 162, 164, 166
human, 4, 7, 8, 9, 10, 11, 32, 36, 38, 39, 40, 53, 55
hydrogen, 43, 46, 49, 52, 58, 62, 75, 76, 77, 79, 98, 99, 100, 101, 102, 103, 105, 106, 110, 113, 114, 115, 128, 129, 141, 142, 162, 164, 166

I

image, 8, 11, 30, 52, 54, 59, 60, 128, 146, 149, 152, 162
imagination, 2, 122, 137
Indians, 11, 12, 26
inertia, 88, 90, 91, 125
interference, 85, 91, 164
international law, 28
intervention, 79, 86, 87, 90, 126
iron, 22, 49
issues, 12, 25, 34, 146, 149

L

laws, 28, 35, 48, 57, 64, 66, 77, 79, 87, 88, 89, 90, 110, 134, 140, 143, 149, 150, 160, 163
light, 3, 10, 30, 38, 39, 40, 41, 44, 45, 46, 50, 52, 53, 60, 70, 73, 74, 77, 83, 87, 91, 92, 125, 126, 137, 141, 154, 163, 164, 166, 167, 169
light beam, 92
lithium, 75

M

Mars, 41, 46, 48, 49, 52
mass, 32, 34, 45, 46, 47, 52, 53, 54, 55, 58, 59, 62, 63, 64, 65, 66, 67, 68, 85, 87, 88, 89, 90, 91, 96, 101, 105, 106, 125, 147, 156, 157, 159, 160, 161, 163, 164, 166, 167
material bodies, , 28, 33, 40, 55, 63, 64, 65, 66, 88, 89, 90, 91, 101, 138, 147, 148, 152, 153, 164
mathematics, 22, 35, 165
matter, 3, 4, 5, 8, 11, 32, 33, 34, 44, 45, 46, 49, 55, 56, 57, 58, 59, 60, 61, 62, 63, 65, 66, 67, 68, 71, 73, 74, 75, 77, 78, 80, 83, 87, 90, 91, 95, 96, 98, 99, 100, 102, 103, 106, 109, 112, 113, 114, 115, 117, 119, 121, 124, 125, 127, 128, 129, 130, 131, 133, 135, 138, 139, 140, 141, 142, 147, 151, 152, 154, 156, 157, 160, 161, 162, 163, 164, 165, 167, 168
measurement, 15, 16, 18, 19, 20, 21, 22, 23, 31, 41, 53, 76, 83
Mercury, 46, 47, 49
mesons, 67
meteorites, 47
microcosmos, 124
molecules, 29, 34, 38, 75, 153

Moon, xvi, 10, 11, 13, 15, 17, 18, 19, 20, 21, 22, 23, 27, 38, 41, 46, 47, 48, 122, 126, 149
mythology, 11, 12

N

natural laws, 88, 89, 90, 91, 105, 135, 138
neutron stars, 45
neutrons, 33, 45, 58, 59, 60, 62, 67, 68, 69, 75, 79, 91, 96, 97, 98, 99, 102, 104, 105, 106, 107, 110, 112, 114, 119, 128, 153, 155, 161, 162, 163, 164, 165, 166
nitrogen, 47, 77
Nobel Prize, 57, 121
nuclei, 33, 44, 45, 58, 67, 68, 70, 74, 75, 78, 79, 91, 97, 98, 99, 100, 102, 106, 110, 114, 119, 121, 138, 153, 154, 162, 164, 165
nucleons, 59, 119

O

orbit, 51, 64, 99, 100, 131, 159
oxidation, 49
oxygen, 49, 77, 162

P

parallel, xvi, 3, 11, 16, 26, 30, 44, 69, 86, 101, 117, 132
photons, xxiv, 65, 67, 68, 69
physical laws, 73, 74
physical phenomena, 11, 12, 63, 73, 149
physical theories, 61, 81, 106, 110, 125, 134, 139, 145, 146, 147, 148, 149, 151, 155, 156, 157, 158, 160, 168
physics, 1, 2, 26, 34, 35, 36, 38, 45, 57, 59, 63, 66, 67, 68, 69, 72, 77, 78, 79, 88, 91, 93, 101, 103, 109, 111, 139, 147, 149, 152, 153, 155, 165, 166, 167, 168, 169
planets, 23, 27, 28, 29, 30, 40, 46, 47, 48, 49, 50, 51, 75, 126, 131, 132, 147, 166
pointons, 54, 61, 62, 63, 69, 83, 84, 85, 86, 87, 88, 89, 90, 91, 93, 94, 95, 96, 97, 99, 102, 103, 104, 105, 106, 107, 110, 114, 116, 118, 119, 125, 128, 130, 140, 147, 153, 156, 157, 159, 161, 162, 163, 164, 167
pulsars, 43, 45

Q

quanta, 34, 72, 125
quantum mechanics, 155
quarks, 33, 55, 59, 60, 61, 62, 67, 68, 74, 75, 78, 79, 91, 93, 96, 97, 99, 102, 103, 104, 105, 110, 114, 119, 128, 129, 130, 138, 153, 155, 156, 159, 162, 163, 164, 165, 166
quasars, 37, 43, 44, 75, 80, 113, 115, 119, 128, 130, 165

R

radiation, 46, 74, 75, 77, 84, 85, 87, 91, 93, 104, 115, 141, 142, 165, 168
reactions, 2, 8, 31, 32, 37, 40, 44, 45, 53, 54, 57, 61, 62, 63, 66, 67, 68, 69, 70, 80, 81, 83, 84, 85, 86, 87, 88, 89, 90, 91, 93, 97, 101, 103, 104, 106, 107, 109, 111, 114, 116, 117, 118, 119, 121, 123, 125, 133, 135, 137, 138, 139, 140, 141, 142, 145, 147, 151, 152, 154, 156, 157, 159, 162, 163, 164, 165, 166
reality, 13, 20, 23, 24, 25, 121, 132
reasoning, 3, 7, 10, 11, 22, 68, 74
relativity, 34, 35, 72, 76, 82, 109, 110, 125, 150, 155, 163, 165, 166, 167

repulsion, 3, 66, 85, 87, 101, 112, 113, 115, 117, 131, 132, 141, 142

S

science, 2, 19, 22, 26, 33, 36, 38, 48, 74, 78, 81, 84, 90, 93, 123, 124, 125, 137, 138, 145, 147, 152, 157, 169
scientific theory, 133
self-destruction, 68, 94, 99, 111
Siberia, 51
solar system, 2, 4, 15, 16, 22, 27, 30, 34, 37, 41, 44, 46, 47, 48, 49, 50, 51, 52, 53, 75, 80, 88, 115, 119, 122, 126, 128, 129, 130, 131, 132, 152, 153, 162, 165, 166
space-time, 76, 150, 166
special theory of relativity, 34
speed of light, 31, 76, 123, 125, 126, 166, 167
spin, 45, 58, 59, 67, 68, 97, 121, 162
stabilization, 100, 102, 103, 119
stars, 2, 4, 9, 10, 11, 14, 15, 22, 23, 25, 26, 30, 34, 37, 38, 39, 40, 41, 42, 43, 44, 45, 46, 50, 51, 52, 53, 64, 75, 80, 88, 101, 115, 119, 127, 128, 130, 147, 152, 153, 162, 163, 164, 165, 166
storms, 47, 48, 50
Sun 9, 10, 11, 12, 13, 14, 15, 16, 19, 20, 21, 23, 24, 25, 27, 28, 29, 37, 40, 41, 42, 45, 46, 47, 48, 49, 50, 51, 52, 64, 67, 148, 149, 166

T

temperature, 39, 40, 46, 47, 49, 50, 69, 72, 73, 74, 75, 77, 78, 79, 80, 114, 116, 119, 166
total energy, 85, 110, 140
Trilogy of the Creation, 1, 2, 36, 45, 66, 68, 98, 104, 110, 112, 114, 145, 147, 154, 155, 157, 159, 160

U

unification, 34, 35, 61, 74, 79, 81, 106, 110, 139, 145, 146, 147, 148, 149, 150, 151, 152, 154, 155, 156, 157, 159, 160, 167, 168, 169
uniform, 77, 106, 145, 165
Upper Force, 12, 86, 87

V

velocity, 30
Venus, 46, 47
Viking, 49

W

water, 10, 12, 16, 48, 56
weak interaction, 111